Postpolitics and the Limits of Nature

SUNY series in New Political Science

Bradley J. Macdonald, editor

Postpolitics and the Limits of Nature

Critical Theory, Moral Authority, and Radicalism in the Anthropocene

ANDY SCERRI

SUNY PRESS

Publication of this book was made possible with financial support from the Faculty Book Publishing Subvention Fund at Virginia Tech.

Published by State University of New York Press, Albany

Printed in the United States of America

For information, contact State University of New York Press, Albany, NY
www.sunypress.edu

Library of Congress Cataloging-in-Publication Data

Names: Scerri, Andy, author.
Title: Postpolitics and the limits of nature : critical theory, moral authority, and radicalism in the anthropocene / Andy Scerri.
Description: Albany, NY : State University of New York Press, [2019] | Series: SUNY series in new political science | Includes bibliographical references and index.
Identifiers: LCCN 2017059916 | ISBN 9781438472133 (hardcover : alk. paper) | ISBN 9781438472157 (ebook)
Subjects: LCSH: Radicalism. | Climatic changes—Political aspects. | Income distribution—Political aspects. | Globalization—Political aspects. | Occupy movement. | Critical theory.
Classification: LCC HN49.R33 S44 2018 | DDC 303.48/4—dc23
LC record available at https://lccn.loc.gov/2017059916

10 9 8 7 6 5 4 3 2 1

In memory of Alan Patrick Roberts

Contents

Acknowledgments

Many people in Australia, Canada, India, the United Kingdom, and the United States helped to shape the argument presented here. Central influences came from longtime mentors, Paul James, Manfred Steger, and Tom Nairn. In this regard, I also thank Tim Luke, University Distinguished Professor of political science, and François Debrix, director of the ASPECT program, at Virginia Tech. I am indebted to Bradley McDonald Michael Rinella, and Jenn Bennett-Genthner for their confidence in my ideas. I thank Karen Hult, chair of the Department of Political Science, for her sound advice, and am grateful for the support of former dean of the College of Liberal Arts and Human Sciences, Elizabeth Spiller, and the current dean, Margaret Blieszner. I am thankful to Debra Stoudt, Janell Watson, Tom Ewing, Peter Porter, and the university subvention committee for their support. I am also extremely thankful to the generations of members of the Caucus for a New Political Science for maintaining a forum for scholarship of the kind to which this book endeavors to contribute.

I am especially grateful to Elisabeth Chaves for her comments on versions of this work, and to Tim Luke, Arran Gare, Meg Holden, Andy Denis, Andy Cornell, and Metin Guven for commenting on all or part of the final draft. I also thank the anonymous reviewers for their many insightful comments and valuable suggestions. I am thankful for comments relating to significant sections of the work-in-progress presented at conferences and seminars, and offered by Andrew Biro, Geoffrey Whitehall, the students at Acadia University's Social and Political Theory Colloquium, Zev Trachtenberg, Peter Cannavò, Jordan Laney, Gabriel Piser, Christian Matheis, Adrian Parr, Jenn Lawrence, Sarah-Marie Wiebe, Emily Howard, Warren Magnusson, Sarah Surak, Yogi Hendlin, and Michael Lipscomb. And, I thank Alex Loftus, Bryan S. Turner, and Jai Sen for their comments on work that appears in revised form here.

I am especially thankful to the many US-based activists and organizers with whom I have spoken and corresponded. Notably, I thank Jonathan Bix and Spencer Resnik at Nobody Leaves Mid-Hudson for their patient and sophisticated account of disenchantment as Occupy activists and reenchantment as highly effective organizers. I thank the many people with whom I spoke at Left Forum in New York City in 2016, Tarence Ray for conversations on organizing in coal country, and Blair Taylor, Ian Welsh, Alex Prichard, and Andy Denis for providing insight into their own work. In addition, I thank Blair for hosting the Institute for Social Ecology Summer Gathering in Vermont in 2016, and Andy Cornell, Brian Tokar, Dan Chodorkoff, Ynestra King, Bea Bookchin, Metin Guven, and Saladdin Ahmed for comments on a paper presented there. I am grateful for the research assistance provided by three former Virginia Tech students, Nader Sobhani, Stephanie Tripodi, and Rebecca Moser. On a different note, I am thankful for the friendship of Tim Parrish, commiserate of the unlucky few, and grateful to Melissa Womble and her team at INOVA Sports Medicine in Fairfax, Virginia who put me back together after an accident that slowed progress in the final stages of research. Most importantly, I am deeply grateful for the unwavering friendship and enduring support of Elisabeth Chaves.

Earlier versions of work in chapters 2, 3, and 5 were published as Andy Scerri, "The World Social Forum: Another World Might Be Possible," *Social Movement Studies* 12, no. 1 (2013): 111–20; "Deep Ecology, the Holistic Critique of Enlightenment Dualism, and the Irony of History," *Environmental Values* 25, no. 5 (2016): 527–51; "Reaganomics," in *Wiley Encyclopedia of Social Theory*, ed. Brian S. Turner, 1939–41, New York: Wiley, 2017; and Andy Scerri and Nader Sobhani, "Even Natural Disasters Are Unlikely to Slow Us Down . . . ," in *Biopolitical Disaster*, eds. Jennifer L. Lawrence and Sara Marie Wiebe, 62–77, Abingdon, UK: Routledge, 2018.

Introduction

The idea that radical and reformist political movements are in crisis has been commonplace since the collapse of the Soviet Bloc in the early 1990s, and more so since the global financial crisis broke over a decade ago. Although signs of renewal have emerged in Western Europe, where some reformist parties have considered abandoning the third way compromise with conservatives, in the United States both radical movements and the reformist wing of the Democratic Party remain vexed. The incompetence and disarray of that party in the lead up to the 2016 presidential election helped to deliver the balance of power in each branch of government to the Republican Party. Meanwhile, radicals began to entertain the possibility that much of the time and energy devoted to prefigurative tactics and ontological politics, which had reached an apogee of incoherence in the Occupy events of 2011, had been for little. Moreover, both radicals and reformists alike came to recognize that they confront well-funded conservative and reactionary movements that for decades have been intent on reshaping politics, policy and culture, as well as the law and the production of knowledge, as extensions of an inegalitarian and in some ways authoritarian market economy.

Both classical Marxists and liberals have addressed political, economic, and cultural manifestations of this contemporary crisis. However, the current crisis also seems to bring to culmination longer-standing tensions within what historians of ideas call the radical Enlightenment project. One body of thought that builds on this radical tradition and which from its inception sought to address tensions internal to the project was the so-called Frankfurt School's neo-Marxist critical theory. Established in Germany in the 1920s, critical theory was carried to the United States in the late 1930s, where it has exerted a profound influence ever since.

Recently, sympathetic political philosopher Raymond Geuss has emphasized how the work of one thinker central to the development of critical theory, Theodor Adorno, addressed the tensions inherent in the radical worldview. In Geuss's view, Adorno uneasily combined the insights of Karl Marx with those of Friedrich Nietzsche to presciently express ideas that would by the late twentieth century become widespread. The collective "loss of meaning" that Adorno sought to elaborate indicates for Geuss a situation in which "people on the Left find increasingly that they have lost faith in the traditional diagnosis or in some part of the traditional recommended therapy" for challenging domination as unfreedom.[1]

Through engagement with work by Geuss, and another Nietzschean political philosopher who was somewhat less sympathetic to both critical theory and Marx, Bernard Williams, I use chapter 1 to revisit Adorno's melancholic insight into the fate that befell the radical Enlightenment aspiration to universalize freedom. Chapter 1 pays particular attention to Horkheimer and Adorno's early philosophical analysis of the transformation of Enlightenment's highest ideal—freedom—into a tool of domination and to the modernist lens that Adorno fashioned in his later work to examine this transformation. Using their work and that of Geuss and Williams, I describe the historical conditions in which actions justified on the basis of the discourse of Enlightenment fostered the emergence of a liberal, democratic, and market-centered form of rationalized authority. And, I set out the theoretical framework for the book in Adornian terms, as a dialectical process that can be observed qualitatively as the critique of injustice under rationalized authority comes to be identified with or assimilated to rationalized authority itself, and subsequently over time to be recuperated as a false kind of non-critique that serves a reactionary role that not only perpetuates unjust dimensions of rationalized authority but threatens such authority itself. Chapter 1 operationalizes this reading of Adorno's dialectic through engagement with Williams's accounts of rationalized political authority as the least worst option of organizing the rules for life held in common. Rationalized authority is justifiable from within both radical and reformist perspectives so long as it responds to claims that injustice is present, even though primarily oriented to sustaining what Thomas Hobbes saw as civil peace and defense against enemies abroad. In the view that I develop in chapter 1, what keeps rationalized authority responsive to such claims is critique. And, loosely following Williams, what threatens to tilt the balance from this least worst to worse kinds of authority are demands couched in terms of moral absolutes.

In chapter 2, I use this framework to examine radical scholarship and activism that, beginning in the late 1960s and early 1970s, drew on early critical theory to elaborate a new critique of domination under the liberal-democratic, military-industrial welfare state. This environmental, ecological, or simply green critique drew on ideas first elaborated in Max Horkheimer and Adorno's 1947 work *Dialectic of Enlightenment*. In particular, radical "deep" and "social" ecologists drew on the critical theory idea of a society run rampant in its efforts to dominate nature. The radical ecologists used this critique to explain to a new generation how long-dominant dualistic Enlightenment values had led humanity to the brink of planetary destruction. Yet deep and social ecology were only two currents within the wave of new social movements to emerge at this time. In this sense, both sit within the context of a broader and deeper tidal turn. In part responding to the shortcomings of left-liberal demands for universal rights, this tide carried radicals in the direction of a resurgent celebration of community. Advocated by philosophers and public intellectuals such as Charles Taylor and Alasdair MacIntyre, the communitarian revival dovetailed neatly with, and in Taylor's case at least interacted closely with, deep and social ecology. Alongside ecological ideas, Taylor's expressive modernism and MacIntyre's call for a return to community posed what was at the time an effective critical challenge to the utilitarian calculus. In the process making "the environment" an object of both critique and authority, the new green-left modernist movements elaborated a way forward for radicals that drew on contemporaneous enthusiasm for community over the state and moral suasion over collective political power.

In chapter 3, I examine the first salvos launched on this new ecological communal terrain for political battles through the lens of what so-called second generation critical theorist Jürgen Habermas labeled at the time a legitimation crisis. In Habermas's view, such a crisis marked an important historical turning point for radicals and reformists. From the late 1960s onwards, government and business had come to realize that for the first time in almost a century neither no longer had any reason to take into account leftist demands. For him, those holding power at the core of the state-market nexus could not fulfill new social movement demands for a post-capitalist system sustained by morally absolute communitarian ideals of holistic ethical harmony with nature without that is, dissolving that nexus. The old modern Left and the modernist new social movements pushed to breaking point the repertoire of social problems to which those holding power would respond. However, Habermas's was not the only intellectual response to the crisis of the mid-1970s. Chapter 3 also discusses the crisis narrative developed at this time by what would emerge as the New Right. Discussion first focuses on the Trilateral Commission's rather sensational report, *The Crisis of Democracy*. I examine how this crisis narrative was taken up and amplified by a host of well-funded scholars and activists in the lead up to the Reagan Revolution of the early 1980s. Advocates of the Virginia school of public choice theory, promoters of law and economics theory, and, importantly, public intellectuals such as Milton Friedman and F. A. Hayek used private sponsorship to argue publicly that not only the radical and reformist Left but also a craven and pandering patrician conservative elite had been complicit in promoting a kind of civilizational hubris. For the Reaganite Right, a return to effective and just authority would require a return to humility and awe at the unfolding cosmological order of things. The New Right's project was grounded in the moral belief that Enlightenment's experiment with the universalization of political freedom had failed. What was conceived as artificial, as unjustifiably grounded in a false dualistic ontology, was intervention by authority that strayed from the task of enforcing the natural Darwinian order of things. Firing an early salvo in the war to "make America great again," these people sought to displace collapsing rationalized authority with the moral authority of the righteous.

Discussion in chapter 4 focuses on the recuperation of radical green-left modernist notions of ecological holism and communitarian modernism to the "New" Democratic Party's reformist electoral platform in the 1990s. The chapter traces the Clinton-Gore campaign's ultimately failed effort to green the market economy through its third way agenda. This agenda was designed to help the Party achieve electoral gains while maintaining a reformist commitment to universalizing conditions for freedom. Accosted by the New Right in Congress and the states from the outset, the New Democrats almost immediately abandoned aspirations to "ecological modernization" and instead embraced the moralized politics of fear. The Clinton-Gore administration would enact a raft of policies that have come to be derided as left neoliberalism. The distinction between such electoral left neoliberalism, the electoral and "radical" right becomes important later in chapter 4, when I shift focus to what I describe as the first generation popular neocommunitarian response to left neoliberalism. At the turn of the century, the George W. Bush administration used such popular neocommunitarianism as part of its "compassionate conservative" policy platform. Against the events of September 11, 2001, the administration pursued a "global governance agenda" and a parallel deregulatory domestic agenda that, as had the third way, emphasized bipartisan support for "corporate social and environmental responsibility" and other forms of civil privatism. Insofar as the left neoliberal eco-modernizing agenda channeled green-left radicals' protests into an efficacious justification for economic growth, popular neocommunitarianism would soften and humanize the pure utilitarian egoism championed during the Reagan Revolution through patriotic appeals to the community of "real Americans."

In chapter 5, I draw attention to what became known as the radical "externalization thesis." Ironically developed by advocates of public choice theory in the mid-1970s in an effort to discredit labor union demands for wage increases in the face of rising inflation as inimical to the greater good, the externalization thesis was radicalized in the 1990s by what the media at the time dubbed a new New Left "global justice movement." The radicalized externalization thesis posited that states were unjustly allowing businesses to externalize the social and environmental costs of production. Of central concern in chapter 5 are electoral party and business responses to this new New Left. In response to the radicals, an organized corporate social and environmental responsibility movement soon took shape. Meanwhile, a slow shift in Supreme Court jurisprudence emerged to endorse corporate personhood, civil privatism, property ownership, and wealth maximization as the most appropriate responses to social and environmental problems. Informed by the public choice doctrine-and-law and economics legal theory, such endorsement helped to depoliticize the externalization thesis by undermining the deontological function of rationalized political authority that has been imposed by the old Left. Having abandoned rationalized authority, the global justice movement soon found itself supporting rather than opposing naturalistic justifications for the expansion of market logic into areas of social life hitherto thought of as external to and operating on a logic different from that of markets. Whereas in the 1970s the radical green-left modernist critique of the centralized, bureaucratic welfare state had combined holistic ecology with moral voluntarism to inform demands for a truly mimetic authority that would express the benign truth that everything is connected to everything else, by the first decade of the new millennium, state and market actors had themselves embraced just such a holistic critique of rationalized authority. By the first decade of the new millennium, it seemed that radicals, reformists, conservatives, and reactionaries alike had accepted that artificially rationalized authority was a failure and should be replaced by some kind of natural ordering force.

The radicalization of the externalization had been at least in part facilitated through engagement with then fashionable ideas of poststructuralism or postmodernism. In chapter 6, I take issue with neoanarchist and radical communitarian efforts to revive ontological politics on the basis of postmodern holistic arguments that seem ultimately driven by Spinozist and Heideggerian nature philosophies. The product was an approach to radical thought and agency that privileged a combination of wild-eyed anti-nomianism and earnest moral suasion over what sociologist Max Weber had famously described as the strong and slow boring of hard boards. I examine how, in the wake of newly elected president Barack Obama's anemic response to large-scale malfeasance on the part of Wall Street, these new radical movements gained significant media attention and, it would seem, public support as representatives of "the 99%." Radicals signally failed to capitalize politically on widespread incredulity at the reformist party's response to the crisis. The ultrahermetic Occupy Wall Street phenomenon, and indeed many of the radical anti-corporate globalization groups associated with it, simply evaporated.

Chapter 6 then identifies a growing mass of scholars and activists who began in the 2010s to question many of the assumptions behind the green-left revival that had begun in the 1990s. The chapter shows that through engagement with civil rights-inspired environmental justice campaigners and with labor unions, tenants' associations, and reformists in an increasingly rudderless Democratic Party, a new wave of green-leftism has begun to take shape. The chapter interprets this new wave of commitment to universalizing freedom as a challenge to outmoded yet still seemingly popular modernist commitments to ontological ecospheric holism and prefigurative moral voluntarism. Chapter 6 construes this new wave of green-left groups that distinguish between moral *activism* and political *organizing* as contributors to the renewal of radical democratic critique. This new-wave radicalism eschews both ultra radical ecological arguments from ontology and irenic liberal everything-is-connected-to-everything-else-in-the-Anthropocene ethics as empty platitudes. Instead, it favors confrontational inside-outside strategies that combine public protest and member empowerment with legal and electoral mobilization. These new wave green-leftists are developing a critique of the holistic critique of Enlightenment dualism and in doing so possibly helping to bring back meaning to Geuss's "people on the Left." Even so, the new groups also represent a certain resignation born of adjustment to the reality that rationalized authority is regrettably the only tool available for resisting resurgent moral authority and so the raw power that Adorno feared would accompany the triumph of moral absolutism over the least-worst primacy of rational principle.

To conclude the book, I summarize the argument and provide a critique of recently influential theories of democratic agonism. Eschewing neoanarchist and radical communitarian efforts to politicize "ontology" and democratic agonism, I argue for an ontological agnosticism that involves regarding authority as simultaneously constituted on two planes: on the one hand, a natural ontological plane—the ecosphere in which human and nonhuman bodies perform, as biological specimens, in space and time—and, on the other hand, an artificial real plane—the realm of institutions that humans, acting collectively, devise and maintain, such that authority is rationalized and as a side-effect may sustain demands for the expansion of political freedom. The implication here is that, should critique continue to be guided by the holistic critique of dualism, it risks remaining mired in apolitical and ineffective moralizing. The alternative to ineffective moralizing of course is effective moralizing, which in the context of a resurgent Right seems to center on a malignant authoritarianism that takes as legitimate only Darwinian principles of natural selection.

Given the increasingly tumultuous political climate in the West, and the United States in particular, since the turn of the century, it does not seem far fetched to see the malignant moralizing central to second generation popular neocommunitarianism as a real threat to the least-worst option for universalizing freedom, the rationalized authority of the liberal, or more ambitiously, social-democratic state.[2] Understood in this light, my own perspective on the effectiveness of what I call radical green-left modernism has altered somewhat since publication of my first book in 2012. *Greening Citizenship* was written as the global wave of Occupy protest flourished. In the ensuing years, having spoken with dozens of radical scholars and activists across the United States and indeed the world, I have become increasingly dissatisfied with the conclusions reached in *Greening Citizenship*. My views on the influence exerted by two of the green-left's core conceptual tools—the holistic critique of Enlightenment dualism and the ethics of moral voluntarism—

have become somewhat jaundiced. In that book, I argued pragmatically that it was entirely possible and desirable for radicals to adapt their concerns with an egalitarian conception of social citizenship to Greens' holistic understanding of society as but one participant in the whole ecosphere. As will become clear, I have become increasingly skeptical of the unabashed ecological holism shared by many radicals and indeed reformists dedicated to green growth capitalism. Today, such holism and its concomitant moral voluntarism seem inadequate as intellectual grounds for interpreting what the Caucus for a New Political Science sees as the struggle for a better world. I hope that you will find this book a hopeful, if provocative, portrait of how meaning was lost, and see it as a contribution to finding meaning once more.

Chapter 1

From Mimetic Expression to the Rational Mastery of Nature

The Discourse of Enlightenment in Early Critical Theory

Political philosophers and political theorists have recently debated the merits of a revival of realist understandings of democratic politics. Although review of these debates stands beyond the scope of this book, this renewed energy for political realism has also served to reinvigorate critical theory. At the center of this revival lay work by Raymond Geuss and Bernard Williams. Both political philosophers decry what they see as the dominance of so-called ethics-first or moralizing approaches to the study of politics since the 1970s, and link this dominance to some of the most telling failures of liberal-democratic ideology and practice over the same period.[1] From the political realist perspective, what is important about interpreting, explaining, and so acting in politically effective ways is not that inquiry begins from a clear and rational idea: an ethical ideal, moral assertion, or I add, a naturalistic ontological truth. Rather, political realism seeks to uncover the justifications that different actors offer for supporting or challenging the use of power by a prevailing authority, and to explain why it is that some succeed and others fail at such tasks. The approach is to begin from plain-language statements that address observable consequences of the exercise of political power in a given historical context. Hence, Geuss and Williams are concerned that theory be not only realistic but also in some way critical of the illegitimate exercise of authority. Indeed, both engage closely with work developed at the Frankfurt Institute for Social Research, the so-called Frankfurt School of critical theory. Of particular interest to both are early iterations of critical theory developed under the practical leadership of Max Horkheimer and the philosophical leadership of Theodor Adorno. Yet, both Geuss and Williams are themselves critical about this body of work. While Geuss reads Adorno's work approvingly, he holds significantly strong reservations concerning more recent iterations of critical theory in a neo-Kantian vein, such as developed by Jürgen Habermas and Axel Honneth, as well as by the avowed liberal John Rawls. For Williams, the problems of critical theory begin with its obfuscatory style of thought and culminate in an erroneous emphasis on freedom instead of the theory of justice. Both Geuss and Williams fault critical theory and especially its more recent highly normative iterations for prioritizing moral concerns with the essence of freedom over political concerns with the relationship between freedom, authority, and its legitimacy.[2] Consequently, both Geuss and Williams adapt early critical theory to develop a political realism that might adequately serve to critique some of the terms on which the dominant justify their authority within the given political order and on which the dominated grant legitimacy to or challenge that authority.

In Geuss's view, while there is "no *single* invariable notion of 'criticism,'" the idea of a realist critique is more often than not inseparable from, and so shaped by, someone's lived experience of domination.[3] For Williams, "the power of coercion offered simply as the power of coercion cannot justify its own use," because for several hundred years, at least, political power has rested on the legitimacy of political authority.[4] In light of their work, a critical political realism would focus on those aspects of the political order that allow some to dominate others in ways that the dominated find unjust. Informed by these views, critical inquiry might ask how and under what circumstances authority acquires certain values, how it comes about that some values trump others, and who is benefited and who is disadvantaged by a political order in which such values circulate. Criticism would be, in the realist view, first and foremost historical, contextualizing and focused on neither solely ideas and values nor actions and practices. Rather, it would focus on the uses and abuses of ideas by actors in the historical conditions that such actors themselves help to constitute and reconstitute over time.

A critical, politically realistic account of the relationship between radical and Green critics of the political order in the United States since the 1970s might therefore begin by recounting the genesis, triumph, and declining influence of one set of values in particular: those associated with the Western European Enlightenment. Of central concern to the strand of radical critique elaborated in critical theory, these values helped to reconfigure traditional ideas concerning the status of society in relation to human and nonhuman nature. They were of central concern because, as a guide to action, they helped to usher in a distinctive and enduring political order. This introductory discussion first uses the realist perspective to reconstruct the critical theory account of the dialectical genesis and triumph of Enlightenment values, paying attention to Horkheimer and Adorno's analysis of the transformation of Enlightenment's highest ideal: freedom.

Enlightenment values helped to foster those bourgeois revolutions—the English, American, and French—through which, for the first time in Western history, since antiquity at least, a significant bloc of individuals believed themselves justified in opposing transcendentally justified moral authority.[5] Defined most famously by Immanuel Kant in the late eighteenth century, Enlightenment prompts persons to autonomously evaluate their interests without direction from another. The "another" in question here is an earthly authority pretending to hold exclusive access to transcendent truth of what is ultimately the cosmological order of things. "Enlightened" individuals act freely, in accordance with their own innate capacities for reason. Secularizing the Judeo-Christian belief in the equality of human souls, Kant held that the human capacity for reason was the equal heritage of all. For Kant, as for John Locke, enlightened individuals base their moral choices not on some given conception of the eternal cosmological order, under advice from religious leaders or powerful elites, but on reasoned reflection, on subjective observation of and action upon the material world.[6] With Enlightenment, individuals can, and should, choose to act according to the dictates of reason. Authority, once conceived as transcendental in origin, manifests as a possibility immanent within each individual. The self-directing individual subject, using reason instrumentally as a guide to effective action, thus operationalizes Enlightenment's highest ideal: the genuine freedom of self-direction, of autonomy in relation to others and the world. As Geuss points out, however, such an ideal of freedom could perhaps only ever amount to a quasi metaphysics, what Williams calls into question as an ethically thin concept.[7]

As Horkheimer and Adorno argue, Enlightenment ideas undermined the highest value and most powerful source of authority in allegedly unenlightened, traditional, or primitive societies.[8] In traditional societies, authority is justified in terms of harmony within the whole, naturally given cosmological order. This is the justification for authority that Horkheimer and Adorno famously describe in anthropological terms as *mimesis* or mimetic immersion within nature. In traditional societies, individuals seek to achieve a mimetic relationship of harmony within the cosmological order. Individuals do this by imitating, copying, replicating, and, importantly, *expressing* natural forces. Politically speaking, public assessment of the quality of the mimetic relationship provides the basis on which shaman, seers, priests, aristocrats, and the like justify elite authority. Elites express natural forces and so maintain the cosmological equilibrium. In Geuss's succinct paraphrase of Horkheimer and Adorno, prior to Enlightenment, "the universe had 'meaning' in itself, as an ontological feature (or perhaps behind it, in the form of Ideas)."[9] Without Enlightenment, there is no distinction between society and nature, outside and inside, only harmony within or disruption of the whole. With Enlightenment, however, the universe no longer appears to individuals as a moral continuum in which they are spatially, temporally, and spiritually immersed, and should feel themselves to be in tune. Individuals no longer experience fate through a value system in which normalcy is predefined. No longer destined by luck or good fortune, individuals apply their own unique yet universally held wills to the world in order to improve it. Effectively freed to master or *repress* nature, individuals are also freed to master the natural passions. Individuals achieve autonomy—and therefore meaning—by making the rational moral choice, and are judged not by their status within a given, eternal hierarchy but by their deeds; their efforts and labors amongst equals.[10]

Horkheimer and Adorno's argument is that, with the spread of Enlightenment values, the traditional decentered source of meaning, and so authority, in society comes to be centered in the modern individual subject. Insofar as modern individuals assert their wills against the world and seek to change it, personhood is liberated from what Kant saw as minority or infancy. With Enlightenment, individuals are no longer justified in blindly conforming to what now appears to be the irrational authority of traditional elites. The holistic interest in maintaining harmony within the cosmological order gives way to a dualistic interest in the domination of nature by a society of "free" individuals. This dualistic distinction—with pure rationality on one transcendent side and individuals confronting choices to concur with it, so to speak, on the other—emerges as the reference point for meaning. What would come to be known as Hobbesian, Cartesian, Newtonian, Promethean, or nature-culture dualism thus grounds a new political and cultural sensibility, which Benjamin Constant defined in the wake of the French Revolution as that of the *moderns*. Politically, enlightened individuals collectively set themselves against the natural order of things to constitute a free society. This impact of Enlightenment values on normative justifications for authority was a consequence of the reduction of what had been hitherto conceived holistically. Experience and understanding of the cosmological order—as opposed to what Geuss recognizes as "the ancient forms of natural philosophy, [in which] the universe had 'meaning' in itself, as an ontological feature (or perhaps behind it, in the form of Ideas)"—was now reduced to a dualistic field of objectively discernible material forces.[12] Enlightenment entails the idea that authority is no longer justified in unreflectively expressing nature. Rather, modern authority is a product of the human capacity to reflect, or alternately to derive principles from, nature. Modern authority may justifiably, as Thomas Hobbes argued, reflect the laws of nature back upon nature to preserve civil peace. Or, in neoclassical terms, modern rational authority may justifiably derive social laws from natural ones, as Aristotle had argued, to promote human flourishing.[11] In both senses, Enlightenment serves authority by justifying the use of reason to subjugate and control nature to advance the human end of freedom.

Politically speaking, modern authority and the law it upholds come to be regarded as the artificial products of an individual-centered order, justice the artifice of collective human design, and freedom the ends shared by both. Yet, the establishment of modern artificial authority justified in terms of reason was the product of historical circumstance, and its establishment was contingent on the actions of identifiable individuals acting in groups, in a particular context. As with any abstract, thin ideal, some groups of individuals more than others can realize freedom and the members of such groups are more likely than others to see the order that upholds a favorable iteration of the ideal as just. One particular group most vigorously championed the notion that authority could instantiate Enlightenment reason. This was the educated stratum involved in the pan-European and subsequently Atlantic expansion of capitalistic markets for goods and services. Members of what Habermas labeled the "bourgeois public sphere" were the white, male, literate, land-owning, and mainly Protestant members of a public that conceived itself as the voice of reason against irrational superstition-based church dogma and passion-fueled aristocratic whim.[13] Informed by factual truth ratified by reasoned observation, this bourgeoisie advocated a form of authority that, designed to concur with reason, would advance freedom. Grounding this secular authority were ontological presumptions concerning nature (the objective temporal and spatial realm of material bodies and forces open to domination through reason) and epistemological presumptions concerning human capacities (the moral freedom to choose a rational ethics over, for example, passions or emotions).[14]

Understood in these terms, the English, American, and French bourgeois revolutions are of central importance to the establishment of modern artificial authority. For Geuss, the revolutionaries' "demand for equality of all citizens before the law, which [stood] in opposition to the feudal regime of privilege . . ." required not only the development of novel political institutions but also capture of the state.[15] The revolutionaries' political demands for a republic (or constitutional monarchy) of free and equal individuals shifted the focus of political disputes over authority and justice. Justifications for authority—which had depended on the demonstration of monarchical power to maintain the deity's well-designed cosmos—came to be grounded in public *legitimation* of the rulers' capacity to respond to individuals' collective will. Rulers of the postrevolutionary state could maintain authority only by enacting laws that were justifiable insofar as they could be represented as mastering nature for the human end of freedom. This shift in the justification for authority meant that the revolutionaries "charged [rulers] with addressing themselves to the interests of the ruled." However, it also meant that, "to discharge the duty of serving the popular interest, rulers to some extent [had to] bear the people's passions."[16] After the revolutions, rulers found that they could legitimately bear the people's passions, and so hold power, only by rationalizing such passions.

Such justification for authority required what Michel Foucault, perhaps most famously, but importantly also Louis Dumont, Albert Hirschman, and Norberto Bobbio, analyzed in the 1970s as a kind of casuistry. The solution to the problem of modern political authority involved the dualistic institutionalization of the modern individual as at once a private person and a public citizen. The private person was liberated to pursue his (at the time) personal passions or choices, notably in relation to religious preference. At the same time, the public citizen bore responsibility to recognize that the capacity to freely make such private choices depended on support for authority, rationally organized to defend private choices as expressions of the public interest in freedom.[17] On the one hand, the enlightened bourgeois citizen deemed the freedom to autonomously pursue private choices necessary. This is because only an autonomous citizenry could ensure that political authority would be exercised rationally. On the other hand, the same enlightened bourgeois citizen deemed rational authority to be in the collective interest. This is because only such a rational authority could guarantee the persistence of citizens' individual freedom of choice over time and amidst an anarchic system of competitive, imperialistic nation-states. That is, those holding modern political authority would need to aggregate the free choices that mattered to maintaining it, and did so by imposing the quasi-metaphysics of the utilitarian rational calculus.

No longer an extension of the cosmological order, naturally justified moral authority gives way to artificially justified political authority: *liberal* because limiting the jurisdiction of political authority to public matters; *democratic* insofar as rule must in some way be legitimated by the people; and *market-centered* because the economic realm of private choices is cut loose from the natural realm of cosmological necessity and refounded on rational authority's capacity to aggregate individuals' utilitarian preferences for pursuing happiness and avoiding pain.[18] Political power and the institutions it upholds are also therefore altered. No longer depending merely on the raw power of monarchical passion and courage, power must also rely on objective analysis and rational administration. Those holding political power must wield authority to settle what Williams casts in a Hobbesian lens as "the first political question" of modernity. Settling this question requires

> the securing of order, protection, safety, trust, and the conditions of cooperation. It is "first" because solving it is the condition of solving, indeed posing, any others. It is not (unhappily) first in the sense that once solved, it never has to be solved again. This is particularly important because a solution to the first [political] question being required *all the time*, it is affected by historical circumstances; it is not a matter of arriving at a solution to the first question at the level of state-of-nature theory and then going on to the rest of the agenda.[19]

Modern artificial political authority is premised on the capacity of those holding power to manage citizenly expectations. Those who hold political power and exert authority must in the very least appear to be respondents seeking advice on how to administer the first political question. And, in the modernity of Western Europe around the time of the bourgeois revolutions, authority managed citizenly expectations by rationalizing citizenly demands upon it. Responding to this question implies that modern political authority need observe what Williams calls a "basic legitimation demand." For Williams, to respond adequately, those holding power must concede something to all of those who do not, individually.[20] Highly important to the analytic schema developed here, however, is Geuss's qualification of the ways that legitimation tends to work in modernity. In Geuss's view, legitimation appears to depend on all citizens, but in fact only depends on those capable of making themselves politically significant and effective in the eyes of those holding authority.[21]

Not neutral in any complete sense, an abstract, formalized, and artificial constitutional system, managed by those controlling the state and so appointing its managers and agents, upholds political authority in modernity. In the language used by Horkheimer and Adorno, mimesis gives way to the state-sponsored rational mastery of nature. Authority is reoriented away from ensuring the natural order of things and to ensuring that the rational mastery of nature is organized in such a way that freedom is realized, albeit in the image of significant and effective blocs of citizens. Indeed, this modernization of traditional moral into modern political authority serves to rationalize freedom itself. For modern political authority to function as legitimate, freedom can no longer remain an abstract ideal. Rather, political authority must administer a peculiar kind of justice that promotes political freedom. Political freedom—the artificially sanctioned status of the public citizen as autonomous in relation to others and, finally, the state—is separated from yet dependent on primitive freedom—the autonomy of the private person to act as if unobstructed by some form of collectively imposed coercion.[22] In this light, modern politics centers on contestation over the constitution of political freedom as a limitation on private freedom, on an artificial distinction between the realm of observable, definable, and legislated social relations and the realm of natural, subjective, and experiential expressions of human being. This said, both political and primitive dimensions of freedom alike are premised on citizenly capacity to influence the state. As feminist scholarship shows, access to both dimensions of freedom are facilitated or curtailed by those controlling political institutions. Such institutions may and have excluded some individuals from political freedom on grounds of gendered, ethnocultural, racial, religious, or class categories, for example.[23] As such, for better and worse, bound to observable demands for legitimacy made by significant and effective blocs of citizens, those exercising authority by controlling the state play the central role in defining what freedom there is to experience at a given moment in modernity.

Modern citizens are the subjects and only potentially the agents of politically rationalized freedom. Political freedom is established, maintained, withdrawn, or extended in response to confrontations over what should be the just order that political authority maintains. To be properly political in modernity is necessarily to challenge or support those controlling the state and so be able to reconstitute the boundaries of political freedom. Political challenges to a given constitution of political freedom therefore require the exercise of collective will on the part of those who see themselves as disadvantaged in some way. Analytically speaking, to reconstitute political freedom, such challenges must successfully modify what Paul-Henri Thiry, Baron d'Holbach, modifying Hobbes's own formulation in view of radical Enlightenment goals, saw as a contingent arrangement amongst collectives representing interest-holding actors. Such public welfarism only appears in a methodologically individualistic lens as being oriented to upholding primitive freedom. In Hobbes' conception, fearing a nasty, brutish, and short life, individual citizens each gratefully but begrudgingly cede authority to an all-powerful sovereign in exchange for the peace that will allow them to freely pursue private goals. The artificial state provides individuals with relief from the state of nature. By evoking the specter of personal injury or loss rather than societal collapse, however, this individualist analytic promotes what Corey Robin calls the moralized politics of fear. Or, in Richard Tuck's view, the incorporation of egoistic utilitarian normative presumptions into the Hobbesian analytic underplays the role played by collective pressure in upholding civil peace and so, political freedom.[24]

In this analysis, some kind of social contract emerges. However, such a public welfare contract is established by sufficiently empowered political collectives—social movements, political parties, and other civic associations, such as professional guilds or labor unions, and commercial lobby groups—that bring to bear claims against political authority on the threat of disrupting the civil peace. While the contract between individuals and political authority is indeed conditional on mutual advantages, it is only through collective power that the contract might be altered. Political freedom tends to work in the interests of effectively organized blocs of citizens. Agnostic in relation to the utilitarian calculating ego, such an analysis emphasizes a politically rationalized politics of fear. Political authority sustains begrudging détente amongst collectives whose agents, envisioning a worse outcome for their clients on observing some contingent resolution to the first political question, contest or accede to stalemate. As will become clear, I believe that, since the 1960s and 1970s in the United States, radical efforts to adapt the holistic critique of Enlightenment dualism to the American experience have tended to evaluate the pathologies of modernity on the basis of a deeply individuated moralized politics of fear. In this respect, the book is motivated by a normative concern to shift the perspective of critique from an analytic that prioritizes the achievement of primitive freedom to one that prioritizes the collective power to coerce political authority to define the contours of political freedom differently.

The Critique of Rational Mastery, Left and Right

The revolutionaries' successful grounding of political freedom in the artificial authority of the state is what gave meaning to the ideologies of Left and Right Although there exist many differences between the two, I focus upon the terms on which each relates to political freedom. The modern Left—born of Enlightenment universalism—is from the beginning confined by a paradoxical relationship to political freedom. One political philosopher to grapple with this paradoxical relationship was Jean-Jacques Rousseau. Notably, Rousseau drew inspiration from the pantheistic understanding of nature developed by the thinker who perhaps did the most to radicalize the Enlightenment project, Baruch de Spinoza. For Rousseau, individuals could freely subject themselves to the political collective on the basis of a truly reasoned understanding of natural forces, which Spinoza had argued were always and everywhere material forces. A collective constituted by such reasonable individuals would in fact express the general will, the generalizable interest of all citizens in freedom. A society organized to express the general will would institute laws sufficient to rescue individuals from alienation and so deliver true freedom.[25]

In dialogue with Rousseau, Karl Marx later tied his conception of the emancipatory generalizable interest to the collective assertion of mastery over nature. In Marx's teleology, the bourgeois revolutions had fostered a false freedom. This is because bourgeois freedom is constrained by the nexus of political power with private property rights. The immediate class interests of the bourgeoisie had perverted the authority vested by the revolutionaries in the state, away from its historical potential to sustain the generalizable interest. Rather, the postrevolutionary state promoted a bourgeois goal: expanding a particular kind of rationalized political freedom through markets that, from the perspective of the generalizable interest, was grossly exploitative. The revolutions made it potentially possible for all to experience a meaningful life in terms unknown in traditional society—through autonomous, self-directed labor. However, the form of political authority that the revolutions had produced simultaneously made it impossible to actually experience a meaningful life in such terms, even for the bourgeoisie itself. For Marx, achieving the basic conditions for meaningful life required a collective political project to take control of the economy and so nature. The aim was to universalize conditions in which individuals could experience the essential freedom that, in Spinozist terms, nature intended, and which for Marx, history demanded. This was the genuine freedom of species being.[26] It is worth quoting Geuss's summation of Marx's view at length:

> The basic modality of . . . collective control must be power over nature and mastery over our productive capacities and economic life, a control exercised through science, technology, and politics. Collective productive activities, Marx concludes, are the kernel of a meaningful life. Furthermore, in a properly constituted economic and political order, the very distinction between instrumental and non-instrumental action can be broken down. . . . In a society in which work and collective social life was sufficiently satisfying, one might think, the very question of the "meaning of life" would not arise. The very fact that this question *does* arise for a particular person in a particular society is a sign that that question for that person (in that society) has no answer. "The meaning of life" ought not to be reified. To know "the meaning of life" does not mean to know any possible discursive answer that can be given to questions about life. Questions ostensibly about

> "the meaning of life" are *really* about whether the social processes are satisfactory or whether certain individuals have a certain capacity or skill, whether they "know *how*" to lead a life of a certain kind, and they exhibit this knowledge in the only way such knowledge *can* be exhibited: by actually leading such a life.[27]

Marx recognized that alienation from the satisfactory state of universal species being was a direct consequence of the fact that the agents of such species being—the organized working class—had not seized control of the state and so organized the economy to that end. Friedrich Engels later developed the teleological argument that once the workers had done so the state would in fact wither. In its place would emerge a global commune organized to measure, predict, order, and ultimately control nature so that all individuals experience noninstrumental species being.

Until the mid- to late twentieth century, the most politically effective iterations of modern leftism sought to wrest control of the state in some way or another to such an end. Indeed, state control was put into practice in various ways: as the complete economic control sought in the Stalinist and Maoist Eastern Bloc; the state-administered mixed market favored by center-left social democratic parties in Western Europe; or the state-managed capitalism practiced by center-left democratic and "labor" parties in the Anglosphere. The aim of controlling the state was to institute an economic order wherein the rational mastery of nature would facilitate the liberation of all individuals. The modern Left sought what Geuss calls a change in the "basic economic structure" that is to be "initiated by political action of a certain type that is directed at giving immediate producers more control over their own activity." Geuss argues that it was Adorno who presciently experienced the looming specter of a "loss of meaning on the Left" in the second half of the twentieth century. For Geuss, the loss of meaning that Adorno experienced, and communicated in his work, gradually took hold amongst "people on the Left [as they found] increasingly that they have lost faith in the traditional diagnosis or in some part of the traditional recommended therapy" for challenging domination as unfreedom.[28] This loss of meaning was slowly undermining the political effectiveness of efforts to universalize freedom by working to build, maintain, and expand institutions of a particular kind, those associated with the state. The desired state institutions were to support the ideal of full autonomy for individuals as citizens, at once legal persons and moral beings, but were not doing so. The vaunted progress towards an eventual poststate authority in which all could experience species being was not taking place.

Particularly at issue for Geuss is Adorno's realization that achievement of the ideal of freedom is in all likelihood not a possible or even a desirable goal to expect of the modern state. Nor, indeed, is that ideal amenable to any enduring expression of collective will as the generalizable interest in freedom. Organizing political power sufficient to universalize an abstraction requires the near infinite extension of rational mastery. Geuss draws two problems from Adorno's work: the problem of the crisis of confidence in the diagnosis, which I interpret as the ontological dimension of the loss of meaning; and the problem of the loss of faith in long-held beliefs about the efficacy of the therapy, which I interpret as a concomitant epistemological and ethical dimension of the loss. On this reading of Geuss's argument, the modern Left sought to collectively operationalize the ontological distinction between nature and culture, the human will and the material world—to exert power over nature—in order to universalize conditions through which individuals could experience a comprehensive freedom. In these attempts to universalize the ideal by collectively exploiting the dualistic distinction between society, as the subject of history, and nature, including human nature, as the object upon which that subject acts, the modern Left, paradoxically, sought to exploit an artificial construction, the state, in order to facilitate a condition understood to be natural, freedom.

In contrast with the modern Left, the Right reacts to left achievements by seeking to use the state and markets to delimit the benefits of modernity to a sample of humanity. The Right too confronts a version of the paradox of freedom, which stems from opposition to universalism. The problem confronted by the Right is that a political order that produces privilege through a market economy requires both an extremely permissive "liberal" attitude to individual autonomy and near-total panoptical authority. In contemporary terms, the disruptive entrepreneurialism essential to economic growth depends on maintenance of a strict political economic order. This market order depends on an authority sufficiently empowered to curtail and channel individual autonomy toward the desired end of economic growth. In defense of such an order, the conservative Right seeks to curtail or limit the achievements of modernization to particular individuals, based on one or some combination of allegedly innate qualities: moral fiber, gender, race, ethnicity, religion, sexual orientation, or nationality, for example. The reactionary Right goes further, and seeks to go further. The reactionary Right rejects that dimension of modern authority, which sustains political freedom in favor of raw power and primitive freedom. For both, some hierarchical scale of entitlement selects the deserving from the undeserving. The audience for people on the right is thus always limited to those whose ostensibly deserved privileges are put at risk by an expansion of political freedom. The quintessential response of both the conservative and reactionary Right to left achievements has therefore been to defend those institutional arrangements that sustain the freedom of the happy few. Right thinkers and activists employ a utopian refusal—pragmatically, in the case of conservatives; principled, in the case of reactionaries—to accept a new contour in the artificial constitution of political freedom.

The Left's paradox and the Right's opposition to universal freedom greatly exercised the advocates of critical theory in the mid-twentieth century. What Horkheimer and Adorno recognized was that efforts to put into practice modern leftism were just as likely as European fascism and Christian democracy, and anglophone liberalism and patrician conservatism, to generate severe pathologies. Importantly, the critical theorists' experiences of exile in the United States led them to recognize that the murderousness associated with Cold War efforts to defend capitalism abroad, combined with propaganda representing techno-scientific achievements as unquestionable signs of progress, had metamorphosed the Enlightenment ideal of universal freedom into little more than "a justification for a pernicious form of equality: the conformist equality of atomized consumers." Adorno, Horkheimer, and, in different senses, Marcuse and Habermas, noticed that even efforts to escape this conformity and to celebrate noninstrumental lifestyles—to align human nature with "natural" nature—were politically deeply problematic. Early critical theory is in this sense the product of an effort to problematize the left commitment to dualism and the presumption that state-organized authority can employ reason to measure, order, predict, and control nature in order to universalize freedom. This problematization was achieved by radicalizing, by pushing to its limit, Enlightenment reason. What such radicalization helped the early critical theorists to reveal was that by the mid-twentieth century, the Enlightenment aspiration to universalize the quasi-metaphysics of freedom itself had in part generated pathological consequences.[29]

For Claus Offe, even though the modern Left had organized itself around the opposition, "social justice and economic security vs. private property and economic power" in the nineteenth and early twentieth centuries, by the mid-twentieth century many had come to regard that same modern Left as having unjustifiably joined an "interventionist and redistributive state that would provide citizens not with liberties, but with rights to resources."[30] The modern Left was contributing to production of the pathologies that Horkheimer and Adorno had earlier exposed. In the name of a circumscribed conception of the generalizable interest, the productivist modern Left was oppressing minorities and riding roughshod over nature. The modern Left had reshaped the contours of political freedom by coercing the state to grant a significant and effective bloc of citizens access to a proportion of the spoils produced by the collective, rationalized exercise of mastery over others and nature. Ignored, underplayed, or even subjected to ridicule were the interests of women, youth, nonwhite citizens, citizens of former colonies, Marx's lumpenproletariat, bohemians, artists, homosexuals, and other outsiders, as was the general interest in protecting or nurturing "the environment."[31]

Those historically privileged enough to exert political power sufficient to alter the constitution of political freedom had shaped the state's response to the first political question. Yet, in moving close to the centers of power under rationalized authority, the demands of the modern Left had in fact been assimilated to power. The sample of the population sufficiently powerful to shape the state's response to the first political question was no longer merely that relatively small clique identified by Habermas as constituting the bourgeois public sphere of the eighteenth and early nineteenth centuries. By the mid-twentieth century, those in control of the centralized, administrative governmental bureaucracy and the heirs of the bourgeoisie, the owners and managers of multinational business corporations, had been forced to make concessions to the modern Left through the power of organized labor. Yet, the political project that had mobilized the modern Left was devalued in the process. Supporting the Cold War military-industrial, Keynesian welfare state, the modern Left had joined markets and the state in promoting the rational freedom of the utilitarian calculating egoist, of the citizen as industrial worker, soldier, housewife, and, importantly, consumer. The modern Left had either shed or in fact never worn the historical mantle of critic in the generalizable interest. Rather, the modern Left had emerged as vanguard of what Adorno and others described empirically as the authoritarian personality.[32]

Drawing on Max Weber's sociology, Horkheimer and Adorno proposed that the ideal of progress towards universal freedom through the rational mastery of nature had come to play the part of an alibi. It justified the exercise of power under modern artificial political authority. For them, the employment of reason by those wielding such authority had disenchanted the critical ideal of freedom by identifying it with systemic requirements. The "dream" of progress through rational mastery had assumed the qualities of a reenchanting myth. Freedom under the sway of this myth offered only the semblance of meaning in an otherwise bereft "administered world." In Williams's précis of the critical theorists' realization, the fact that the modern Left could no longer be conceived as avatar of the generalizable interest had made clear that the Enlightenment "aspiration to social management as applied scientific truth and its fantasies of reconstructing human and social relations in a radically rationalistic spirit" were in fact "dangerous delusions."[33] The alibi or myth of progress through rational mastery had truncated the truly liberatory potential of Enlightenment reason. The satisfaction of modern Left critique through the application of instrumental rationality—the assimilation or identification of hitherto radical demands, through the redistribution of a share in the spoils of nature's exploitation to a sample of the working class—had transformed the ideal of a state geared to universalizing freedom into a justification for totalizing authority. This dialectical assimilation subsequently transformed the modern Left's original critique into an unreflexive commitment to truth as essentially, the egoistic utilitarian calculus of costs versus benefits. In winning concessions from those in charge of the system, the modern Left had abandoned the truthfulness of critique mobilized by the experience of unfreedom. As will be shown, perhaps more importantly was that this process helped to recuperate irrationalist romantic, nonidentitarian challenges to rationalized authority. The dialectic of nonidentitarian critique, its assimilation to positive identity, and recuperation as false nonidentitarian reaction, is important. As Williams points out, while the critical, nonidentitarian commitment to truthfulness is speculative, negative, and therefore necessarily implies a "theory of error,"[34] commitments to positive identitarian and false nonidentitarian understandings of truth as essence both imply moral absolutism.

Rationalized Authority and Adorno's Modernism

Setting aside the Marxist heritage, Adorno also owed a particular debt to Friedrich Nietzsche's emphasis on the relativity of values that emerges under the Enlightenment's secular ideal of freedom as autonomous self-direction.[35] In Nietzsche's naturalistic philosophy, in the absence of some cosmologically ordained immutable truth, justifications for authority are relativized. However, recognizing this is not to advocate the sophomoric view that equates relativism with meaninglessness. Rather it is to recognize that justifications for authority come to depend on—are made relative to—different terms at different points in history and in different societies.[36] Following Williams, then, the political climate in the United States in the mid-twentieth century cannot clearly be understood as approximating some benign, pluralistic optimum equilibrium state. Nor can the situation be adequately understood in the faux-Nietzschean terms that a naive reading of Horkheimer and Adorno's thesis may suggest, as the complete annihilation of possibilities for true freedom in the chaos of Darwinian struggle. The situation is better understood as a tense and endless standoff between powerful factions, resulting in a particular constitution of political freedom that defined the public welfare in specific, if overall unsatisfactory, terms.

In the mid-twentieth century, "big government" demanded resources to fuel Cold War militarism, "big business" demanded profits and rents, and the modern Left demanded a high standard of living. In this view, the price of the modern Left's compromise with authority was fealty to an essential truth, that of the egoistic rational calculator motivated to act in terms circumscribed by the moralized politics of fear.

At issue is that the egoistic instrumentalism central to such utilitarianism is prepolitical. The utilitarian ideology of the egoistic rational calculator justifies what Williams calls theorizing on the model of "enactment moralizing":

> The model is that political theory formulates principles, concepts, ideas, and values; and politics (insofar as it does what the theory wants) seeks to express these in political action through persuasion, the use of power and so forth.
>
> The paradigm of theory that implies the enactment model is Utilitarianism. Unless it takes its discredited Invisible Hand form (under which there is nothing for politics to do except to get out of the way and get other people out of the way), this also presents a very clear version of something always implicit in the enactment model, the panoptical view [which means that] the perspective on society is that of surveying it to see how it may be made better.[37]

The theorist or ideologist of utilitarianism favors an apolitical, panoptic conception of authority's relationship to freedom. In terms of practical reasoning, freedom is defined as the happiness that always and everywhere accrues to the rational egoistic calculator of costs against benefits. The task of authority is to re-form individuals in the image of rational authority. Those holding such authority charge themselves with ensuring the greatest happiness of the greatest number. In this way, the ideologist posits that a society organized along rational lines will always and everywhere serve as the most efficient vehicle to ensure deserving individuals, and so ensure national triumph in nature's Darwinian struggle.

As Adorno saw, the problem for the modern Left is that rationalizing freedom as the choice to maximize pleasure and avoid pain assumed away possibilities that the authority charged with upholding freedom through such means may warrant critique from those disadvantaged in some manner. The utilitarian calculus that had once allowed authority to legitimately bear the people's passions no longer represented freedom for significant and politically effective sections of the citizenry. The class whose desires for freedom, mattered to authority was no longer merely the bourgeoisie, who expressed such desires through the pursuit of private property and economic power. The casuistry that had allowed postrevolutionary elites to respond to these demands of the ruled now allowed those holding power to respond to the modern Left's demands for social justice and economic security. As the economy had grown, full citizenship had been extended to wider segments of the population. Yet, with the incorporation of those formerly outside, the working class, the remit of political freedom was further rationalized. Support for freedom in its utilitarian form emerged as anti-political support for the essential truth of the egoistic calculus. As an ideological trope, enactment moralizing represents the interests of the privileged, the lucky few who benefit from the state's response to the first political question. That is, those who benefit from the displacement and marginalization of critique as the negative commitment to truthfulness by critique as the positive commitment to truth as essence.

In light of this account of enactment moralizing as ideology, another dimension of Adorno's debt to Nietzsche becomes important. After all, it was Nietzsche and not Marx who in the late nineteenth century had argued that Enlightenment sustained a kind of civilizational hubris in relation to the natural order. Nietzsche's response to Enlightenment was not to seek collective emancipation by championing such hubris, as did

Marx. Rather, Nietzsche confronts modernity with a newly dawning alternative, a more-than-modern "modernist" value system anchored by the autonomous individual. Uneasily synthesizing Marx with Nietzsche, Adorno seeks to couple the radicalization of Enlightenment as a collective project with a thorough modernism. Adorno draws Marx's concept of the species being from which humanity has been alienated into engagement with Nietzsche's view that modernity entails the relativizing, the transvaluation, of all values. In light of the pathologies that collective efforts to rationally master human and nonhuman nature are bound to generate, Adorno posits a modernism that privileges autonomous self-realization within the otherwise pathological condition that is modernity. As a more-than-modern permutation of Enlightenment, his left modernism renders politics secondary to philosophical insight. The upshot is that critique itself, as Adorno well understood, is removed from the realm of politics, the realm of opportunities to collectively influence authority's response to the first political question in order to expand and enhance the remit of freedom.[38]

In contrast with his colleague and friend Horkheimer's turn to a-religious theology, Adorno unflinchingly embraced the full implications of this Left modernism. His *Negative Dialectics*, *Minima Moralia*, and *Aesthetic Theory* can all be read as addressing directly the left modernist commitment to philosophical reflection on the pathologies of modernity. In these works, Adorno offers little or no means of support for a political project, be it state capture or abolition. As such, he restricts his modernism to the realm of aesthetic evaluation of the awe-inspiring qualities of "the sublime." And, accordingly, he advocates a modernist idea of freedom as the product of truthfulness confronting the sublime. Adorno's modernism seeks to confront the frightening pathologies *and* the liberating opportunities (together, the sublime) that "the system" produces. Central to Adorno's modernism is his conviction that the totally administered world of the mid-twentieth century depends for its reproduction on the dangerously deluded commitment to instrumental rationality, to positive truth as essence.[39] Adorno most clearly recognized two things. First, the world simply did not conform to the dualistic Enlightenment ontology that structured the mid-twentieth century world that he inhabited. Most relevant in this respect is his infamously gnomic pronouncement that "The whole is the false" or alternately, the untrue (*Das Ganze ist das Unwahre*). What I take Adorno to be implying is that humanity does, in fact, participate in nature, in the ecosphere, in a cosmological continuum that extends backwards and forwards in time and across space, and must be understood as such. The dualistic ontology was a dangerous delusion. The aspiration to rational mastery of nature is a pathology-inducing myth. Yet, Adorno also recognizes that the philosophical realization of this fact itself can only take one so far: "Wrong life cannot be lived rightly," or, there is no right life in the wrong one (*Es gibt kein richtiges Leben im falschen*). Therefore, second, his Left modernism allows no grounds for analyzing or normatively promoting collective organization, "The detached observer is as much entangled as the active participant; the only advantage of the former is insight into his entanglement, and the infinitesimal freedom that lies in knowledge as such."[40] The negative commitment to truthfulness is all that philosophical reflection can sustain. He offers a formula for understanding and enduring modernity, not for acting to bring to life an alternative system. The freedom that accompanies the commitment to truthfulness and that allows the sensitive to confront the sublime is not transferable as a guide to collective engagement with political authority.

For Adorno, the loss of meaning that he associated with the radical Enlightenment project was not so much a consequence of the triumph of Enlightenment values over traditional mimesis. Rather, it was located in recognition that a full critique of injustice as unfreedom could only be grounded in a negative response to real world conditions. Insofar as he conceived mimesis as a normative orientation to the world, Adorno himself at times seems to have sought mimetic immersion within a benignly conceived cosmological order. His position within that order tended to be one of delivering a verdict on humanity from the perspective of a platonic view from nowhere. Although somewhat conflicted on this issue, Adorno seemed to rue the fact that the only possibility for collectively upholding freedom in modernity—action oriented to altering or expanding the remit of political freedom—fell beyond his negative dialectics.[41] At some point, the effort to universalize freedom necessarily involves engagement in the very un-philosophical, indeed unnatural and so artificial, realm of politics, of taking sides against those who would deny universalism, of building alliances to confront those exercising authority in favor of particularism, of contributing to the job of ensuring a just redistribution of the spoils of humanity's exploitation of nature. As Offe and Brunkhorst suggest, something that Adorno, lifelong supporter of the West German Social Democratic Party, observer of the rule of law, redistributive economic policy, the granting of rights, and publicly funded civic education, failed to emphasize in the early or even middle stages of his career was the acute dependence of the radical Enlightenment project on the cracks and fissures that collective democratic action constantly open up in the system. What Adorno may not have expressed so clearly was the fact that his nonidentitarian capacity for critique, for freedom as the capacity for truthfulness in face of the sublime, depends not on philosophical rigor but on political freedom. Indeed, Offe does argue that Adorno did late in his career begrudginly recognize the resistance of "parliamentary democracy [and] civil society to totalitarian currents," and his "American" work has more recently been interpreted by Shannon Mariotti as a direct contribution to democratic theory.[42]

Moral Absolutism and Mimetic Regression

Adorno no doubt rejected the utilitarian instrumentalism of the truncated Enlightenment. But he also rejected the regressive atavism that he associated with the false yearning for mimetic immersion. That is, his modernism not only rejects instrumentalism but also simultaneously rejects what he sees as ultimately reactionary "mimetic regression" to worship of the natural cosmological order of things. Coinciding with critical theory's first wave of influence in the United States, its foremost thinker elaborated an approach to modernity that rejected both the instrumental utilitarian domination of nature and what by the 1960s had reemerged as an increasingly popular yearning for a return to the authority of mimesis, of an order that *expresses* nature. Indeed, Adorno regarded mimetic atavism as the leitmotif of other Continental philosophers enjoying resurgent influence in the post-1945 era, notably Martin Heidegger. Adorno regarded the mimetic yearning of Heidegger and his followers as fundamentally regressive because it entailed uncritical commitment to the inalienably essential truth allegedly located in the cosmological order itself, the truth of being as presence (Dasein). For Adorno, Heidegger's "mimetic regression" does not directly entail the identitarian thinking of the utilitarian calculating ego, which in assimilating all to "the concept" seeks to dominate blind nature through *repression*. Rather, Heidegger's mimetic regression involves false nonidentitarian thinking. Similar to the original "axioms and definitions of Spinoza," Heidegger's false nonidentities entail "systems thinking" and seek to imitate blind nature by *expressing* it. As responses to modernity, both identity and false nonidentity thinking renounce the radical commitment to truthfulness. What is favored is some kind of positive truth as essence, be it arrived at through dualistic, panoptical enactment moralizing knowledge, oriented to the enforcement of utilitarian calculation, or holistic knowledge of unity with oneness, oriented to re-structuring the world in the image of the ethical knower. Indeed, Adorno saw the latter approach as embodied in Heidegger's account of the path in the woods that guides the noble "founder" of being as presence.[43]

If triumphant, the product of false nonidentity thinking would be for Adorno a "horror." An authority in control of rational means that pretends to express the diffuse forces of nature positions what is ultimately the moral authority of the most willing beyond critique. When the exercise of power under authority pretends to express authentic nature, the cosmological order, critique is once more rendered heretic, as it had been in premodern society. When authority expresses natural forces, protest against injustice is cast as subversion of the natural order of things. Authority would refer not to rational proofs, to evidence, but to the immutable facticity of nature, that is, to some malignant conception of nature as a competitive struggle or, alternately, to some benign conception of nature as a harmonious haven. Authority would refer to the pious moral altruist, cooperative coproducer of harmonious ethical community.

Adorno's distinction between instrumental rationality, mimetic regression, and his own enlightened and, so, "bitter" modernism might be operationalized for political theory by once more calling upon Williams's Nietzschean account of moralizing about politics. Indeed, insofar as Adorno was himself committed to some concept of progress, I believe that this step away from philosophy, and through Williams's vindicatory Nietzschean political philosophy to political theory, is a reasonable move.[44] Williams argues that the legacy bequeathed by Enlightenment to Western modernity, and the United States in particular, sustains an alternative to instrumental enactment moralizing. This similarly moralistic and so apolitical alternative he labeled "structural moralizing." What is important for Williams is that, like its enactment-moralizing counterpart, structural moralizing is also, in effect, a commitment to truth as essence. This is because practical reasoning on the political philosophical model of structural moralizing "lays down moral conditions of co-existence under power, conditions under which power can be justly exercised." Citing the work of John Rawls as "the paradigm of such a theory," Williams argues further that, "under the structural model, morality offers constraints (in

TJ very severe constraints) on what politics can rightfully do [*TJ* refers to Rawls's *Theory of Justice*]."[45] As philosophy or ideology, practical reasoning on the model of structural moralizing aims to limit authority to inalienable and so prepolitically defined pure moral principle, what should count as rightful political action. The objective of structural moralizing is to specify the right, and to hold it up as the definitive constraint on the exercise of authority. Some essential, inalienable truth (in the case at hand, the harmonious equilibrium that local communities constituted by moral volunteers are said to propagate) is taken as the basis for remaking, or indeed purifying, political authority.

Both the instrumental yearning for total enactment and the regressive mimetic yearning for structural oneness undermine rational argument and formal procedure, which Adorno somewhat ambiguously saw as normatively desirable features of modernity. With mimetic regression, the normatively desirable best case at hand is subsumed to the demonstration of private moral authority parading as the expression of immutable forces. False nonidentity thinking is false because it represents human power as if it were the expression of "nature" itself, rather than an instantiation of moral authority. For Adorno, the flawed and dangerous artificial political authority that he identifies with the totally administered world is literally the least worst of possible worlds. Indeed, both Horkheimer and Adorno recognized that "freedom cannot be separated from Enlightenment" and, more pointedly, that "[d]omination, in becoming reified as law and organization has had to limit itself."[46] Therefore, with hindsight, it might be argued that what is at stake for Adorno is a pathology-producing system—the least worst option—that depends for its reproduction on the fact that it contains cracks and fissures, in which immanent negative critique of those pathologies might emerge. Adorno saw that both instrumental utilitarian, identitarian thinking and the mimetic yearning of false nonidentity seek to occlude the often barely discernible or accessible cracks and fissures that exist even though "Society has become the total functional context which liberalism used to think it was." In place of both the positive truth of truncated reason and the cracks and fissures within it that allow for radical truthfulness, false nonidentity thinking poses a holistic moral system: "Its goal, the most ancient of goals, is not truth but absolute semblance, dull imprisonment in a nature we do not see through, a mere parody of the supernatural."[47] Adorno's insight is that once critique abandons the negative effort to confront domination—"the speculative moment" of truthfulness—and instead assumes the positive task of making things better—"positive negation"—the negative commitment morphs dialectically into a positive commitment to truth as essence.[48]

Indeed, it is well known that Adorno expressed surprise and consternation in the 1960s, as New Left and new social movement thinkers and activists in Germany and the United States took up and ran with many of the ideas developed in critical theory, in particular the idea of nature and of the myth of Enlightenment reason.[49] These young radicals, many of whom were the beneficiaries of the full citizenship afforded by the modern Left's compromise with the state and markets, elaborated ideas drawn from critical theory to audiences beyond the academy and within a political culture that had experienced decades of unmitigated economic growth and accessed higher education in numbers greater than ever before. Witnesses to the new radicalism, both Offe and Habermas illustrate the terms on which the New Left and new social movements had begun by the late 1960s to displace the concerns of the old modern Left. The new social movements had subsumed the modern Left's homogeneous concern with economic redistribution from capital to labor to a heterogeneous "scattered set of issues and the incoherent expression of complaints, frustrations, and demands which [appear to] not add up—either ideologically or, for that reason, organizationally—to a unified force or vision" expressing the generalizable interest of all citizens. What made the new social movements, and in particular the Greens, new for Offe was that "the 'enemy' which is to be overcome is no social class or category of people, but some more abstract kind of dominant rationality in which, at least to some extent, 'all of us' do actually partake or upon which we depend." Whereas the bourgeois revolutionaries sought to undermine the hierarchical regime of inherited privilege, and the modern Left the market regime of private property and economic power, the new social movements sought something diffuse, indeed holistic and in no small way closely resembling Adorno's Left modernism. Inasmuch as they could be categorized, Offe argued that the new social movements sought to undermine the values that upheld a world in which "fear, pain, and (physical or symbolic) destruction" proliferated, and uphold those of a new world in which "integrity, recognition, and respect" for social and ecological harmony prevailed.[50] For Habermas, "freed from immediate economic compulsion," those constituting the new social movements demonstrate a

> lack of understanding in principle for the reproduction of virtues and sacrifices that have become superfluous—[and a] lack of understanding why, despite the advanced stage of technological development, the life of the individual is still determined by the dictates of professional careers, the ethics of status competition, and by values of possessive individualism . . . why the institutionalized struggle for existence, the discipline of alienated labor, and the eradication of sensuality and aesthetic gratification are perpetuated.[51]

Although dismayed by and in disagreement with the protesters, it was Adorno who most keenly elaborated the general mood that something was wrong with the modern Left's worldview and with the bureaucratized, centralized, industrial welfare state. Adorno had seen that the key to understanding what was wrong was to be found in some kind of modernism that rejected in principle the reproduction of modern virtues and sacrifices.

Returning to Geuss's thesis on the loss of meaning on the Left, the modern Left had once sought to implement a "change in the basic economic structure initiated by political action of a certain type that is directed at giving immediate producers more control over their own activity." However, at some time in the mid- to late twentieth century, Adorno had distilled into philosophical insight the wider recognition amongst "people on the Left" of a

> lost faith in the traditional diagnosis or in some part of the traditional recommended therapy [for challenging domination]. Either the malaise is not located in the economic structure, but is even more deep-seated, such as in the structure of rationality itself, or the form of political action traditionally recommended by those on the left is likely to be ineffective or counterproductive.[52]

The modern Left's political project lost meaning as it became clear to increasing numbers of people, especially young people, that "the structure of rationality itself" did not conform to the Enlightenment ideal. At the level of ontology, a "loss of meaning" took shape as people came to accept that society and nature are not dualistically separated. Ironically, in part this realization had been made possible by transformations in scientific knowledge of a magnitude at least comparable with those of Enlightenment. Notably physics, biology, and the "new science" of ecology had demonstrated that humanity was no longer objectively the master of nature but, rather, that society was but one participant in the encompassing ecosphere. At the level of epistemology, this loss of meaning took shape as people committed to the radicalization of Enlightenment ideas lost confidence in the ethos of solidarity oriented to collective control. People such as Adorno came to realize that the basic economic structure could be changed through political action to give immediate producers more control over their own activity but that doing this creates further, sometimes new, pathologies. The modern Left had bought into the dualistic reenchanting myth of progress through rational mastery. The egoistic utilitarian calculator was no more than a possessive authoritarian echo of the once emancipatory demand for liberation as autonomy in relation to transcendent authority, and instrumental reason merely the alibi for repressing such autonomy, oppressing dissent, and facilitating the limitless exploitation of nature.

Critique in the Anthropocene "Age of Ecology"

One thinker to highlight the mixed consequences of radical green efforts to harness Left modernism for political ends since the late 1960s is environmental historian Joachim Radkau. Radkau's substantive thesis is that, beginning with the "first great eco-boom" that kicked off around the inaugural Earth Day in the United States in 1970 and extending to "the second great eco-boom" that he argues began in 1990 and was amplified in the wake of the global Earth Summit in 1992, radical Greens would provide nothing less than the motive force behind an "epochal shift." In his view, Greens' central assertion—that society is rightly understood holistically as but one participant in the ecosphere and that individuals should choose to act in ways that render this participation "sustainable"—would undermine the reenchanting myth of Enlightenment and usher in a new Enlightenment, an age of ecology.[53]

Radkau is not alone in identifying the establishment of a new age of ecology in the second half of the twentieth century.[54] Perhaps more well known in the United States is environmental historian Donald Worster. Worster argued in the mid-1970s that the age of ecology actually began in the United States with detonation of the first atomic bomb in New Mexico in 1945. Radkau agrees with Worster that this event may have "marked an epochal change in material reality."[55] As might be debated ad infinitum today, it provides yet another possible commencement date for the Anthropocene.[56] However, Radkau rejects Worster's thesis for being too focused on material incidents and facts. In so doing, Worster fails to account for how such incidents and facts intersect with political culture, namely, with ideology. In contrast, Radkau argues that even traditional societies undertake ecospherically disruptive acts in search of minerals, plants, and animals. The development of nuclear fission was merely the latest in a long history of literally Earth-changing achievements. In Radkau's view, the age of ecology did not begin until people organized large-scale political and cultural protest against the impingements upon freedom that accompanied such Promethean efforts to harness atomic power.

More recent accounts of a nascent age of ecology emphasize how the fact of the Anthropocene might be put to use as mobilizer of a new emancipatory politics. For legal scholar Jedediah Purdy, the Anthropocene signals a welcome end to what he sees as a once progressive but now politically disempowering idea of nature. Purdy argues against the view that nature should be esteemed as a pristine "environment" to be conserved, and for the politicization of society's immersion in nature. Yet, Purdy's advocacy draws on the currently fashionable mélange of ideas falling under the banner of post-humanism, political ontology, the new vitalism, or new materialism. The result is an irenic call for a more intuitive and collaborative autopoietic response to the "Anthropocene question—what kind of world to make together," a response that would "build a peacefully humane world" on the basis of post-human "fear and love."[57] From a critical perspective, Purdy's autopoeitic reflections all too easily evoke the majestic plural, rolling rich and poor, powerful and powerless together into a politically sterile view from nowhere.

In contrast with both Worster's material facts and Purdy's hand wringing, Radkau uses insights drawn from Nietzsche, Weber, and critical theory. He traces the beginnings of the age of ecology to an epochal change in spirit, in human consciousness that was led by the radical green-left modernist movements that arose in the United States and the broader West in the wake of the first Earth Day in 1970.[58] More important than material incidents or some benign autopoietic impulse to equilibrium is his observation that the radical green-left effectively translated the findings of scientific ecology through the perspective offered by the holistic critique of Enlightenment dualism, and did so for generations of political and cultural actors. The radical green-left modernist movements would help to prompt a deep-seated ideological shift by distilling the mood of protest against the modern culture of control into a near universally appealing message: man can no longer legitimately be conceived in Enlightenment terms as master of nature.

I am unaware of contemporaneous criticism of what I will call the *green-left modernist synthesis* that emerged in the late 1960s and early 1970s, other than that posed by Hans Magnus Enzensberger. Enzensberger foresaw clearly in the early 1970s how the moralizing potential inherent in radical green-left modernist pretensions to an ecological science-based political movement might prove problematic in the historical struggle to universalize freedom. Enzensberger argued that green-left modernist efforts to embellish a holistic ecological critique of the pathologies of Enlightenment dualism were bound to foster not the triumph of pure inalienable green truth but rather a new worldly terrain on which political battles would be waged.[59] A little later in the 1970s, Timothy Luke would use Adorno's take on critical theory to highlight the imbrication of radical green-left demands into established market political and economic imperatives. In Luke's view, politicians and business elites would eventually respond to the science-based dimensions of radical demands for a political economy that accounts for the environment, nature, or the planet. In doing so, radical demands would eventually help to sustain a proliferation of numerical models attempting to capture feedbacks between society and the environment. In the world that would be produced, the words and actions that would make the most sense, that would be most rational, would be those couched in terms of holistic ecology.[60]

As does Radkau, Luke argues that this new ecological regime would become global in two senses. First, it is always and everywhere planetary. Insofar as it is known that society's impacts on the environment take place on a global scale, this knowledge integrates all of humanity into a single ecosphere of existence. Following, scientific environmentalism would offer both political rulers and the ruled, sovereigns and subjects, a universal grammar that affords all with opportunities to appear to each other to be acting rationally, thinking right thoughts, and doing good deeds. The combined effect of decades of green-left and subsequently mainstream professional Green movement interventions in politics would be the emergence of politicians eager to please voters with cautious policy and businesses eager to supply consumers with environmentally friendly products. At once material and ideological, for Luke the establishment of the environment as a discursive object, an Adornian constellation, would produce rational actors and impose a green rationale for political decision-making that readily admits, and in fact requires, positive negation, "artificial negativity."[61]

Indeed, Radkau argues with hindsight that the "global rhetoric that characterized the 'ecological revolution' from the first great eco-boom of the 1970s betrays a dream of power beneath the surface:"[62] unwittingly, the new radical green-left movements were helping to establish what Enzensberger and Luke foresaw as a new political terrain on which future conflict between Left universalist and Right particularist ideologies would take place. Just as the Horkheimer and Adorno had realized that by the mid-twentieth century, "Enlightenment became entangled in the mechanisms of power, producing its own kind of intolerance and turning its highest ideal, Reason, into an instrument of domination," so eventually would the holistic critique of Enlightenment dualism itself become a barrier to critique. For Radkau,

> the radical green movement, and its central conception of a holistic ecology today stands in a relationship of dialectical tension to developments in the real world but is so far the only answer to them, or anyway the only one on a scale larger than all the promises of liberalization and globalization. In this sense, it seems justified to speak of an ecological age . . . eco-age [or] New Enlightenment [that first took shape in the 1970s].[63]

Although inspired by Radkau's historical claim that radical green ideas and agency have exerted an epoch-defining ideological impact, I disagree with the implications that he draws from such success. In contrast, I find that the combination of a naturalistic holist ontology with a moral voluntarist epistemology have been so thoroughly assimilated to the ideology of liberalization and globalization that these cannot offer the only answer to real world problems of social and environmental exploitation.[64]

To these ends, work by Ingolfur Blühdorn is also helpful, even though his focus is almost exclusively on Western Europe. Blühdorn develops his analysis of radical Green achievements on the basis of Niklas Luhmann's systems theory, which, as Brunkhorst argues, closely resembles Adorno's effort to unflinchingly confront the sublime that is "modernity." For Blühdorn, the environmentalist, communitarian, and modernist participatory-democratic values extolled by green-left activists in the 1970s would by the 1990s be redeployed as justifications for decentralized, stakeholder-engaging, and "sustainable" forms of governance. As do both Enzensberger and Luke, Blühdorn argues that this transformation of critique has meant that authority would no longer speaks to politically organized, solidary blocs of citizens. A new form of authority would emerge, justified in terms of its capacity to morally motivate stakeholders who would no longer seek to challenge the system but to unquestioningly realize themselves within it. On both sides of the Atlantic, the coercive role once played for better and worse by the workers in deciding the distribution of the spoils of nature's exploitation would be displaced by citizen volunteers ostensibly choosing between ethical and unethical goods and services. What the green-left radicals in the late 1960s and early 1970s saw as the desired means for undoing the pathologies of modernity would over the ensuing decades come to define authority's promises. Hitherto radical demands for decentralization, as well as for local community empowerment and individual ethical self-realization would come to justify the postpolitical exercise of moral authority. Such authority would focus on self-actualizing stakeholders, active in their locally sustainable or alternately patriotic heartland communities, but eschewing real politics. That is, the individual *and* the system would come to be constituted and governed by the same naturalistic logic. Authority would become postecological, effectively postdemocratic, and so postpolitical.[65] Framed by Adorno's philosophical distinction between identity, false nonidentity and nonidentity thinking, and Williams's arguably parallel distinction between enactment and structural moralizing and political realism, the following chapter 2 picks up the narrative threads that Radkau, Enzensberger, Luke, and Blühdorn lay out.

Chapter 2

Holism, Modernism, and "the Problem of the Environment"

Everything Is Connected to Everything Else: Deep and Social Ecology

Beginning in the late 1960s, many dispirited New Left and civil rights activists sought to merge a radical political agenda with newly publicized insights from ecological science. In doing so, these scholars and activists helped to consolidate a somewhat novel political movement around the environment. This green movement gained widespread support amongst the demographically significant generation of so-called baby boomers. Perhaps the two most sophisticated efforts to synthesize political radicalism with ecological science were those undertaken by advocates of deep and social ecology. Deep and social ecologists both explicitly sought to distill insights into the pathologies of modernity by drawing directly from critical theory and to elaborate from it an emancipatory political project. Both romantic-idealist-deep and dialectical-materialist-social ecologists justified modernism as a sensibility that concurred with the findings of scientific ecology. Deep and social ecologists sought to reconfigure society not on false dualistic terms, as separate from and master over nature, but rather on benignly whole terms, as but one participant within the ecosphere. And, both pioneered paradigmatically new and enduring intentional, prefigurative, communalist forms of activism.[1]

Unforeseen by many of these scholar-activists at the time, however, would be the wider significance of this turn to community. Over time, wider currents would carry radicalism away from longstanding concerns with the state and capitalism, structure over agency, and constraint and order over unboundedness and disruption. That is, the turn to prefiguring desired future social arrangements in present local communities would distance radicalism from political concerns with authority, legitimacy, and contest, and preoccupy many with concerns over moral choices and ethical intentions. Indeed, what would take shape in the 1970s as green-left modernism both echoed and contributed to a deeper transformation of ideology and political culture in the United States. In some respects, this transformation arose organically out of the countercultural phenomenon of the 1960s. However, the key motifs of community, locale, moral choice, and ethical concern would later be actively encouraged and sponsored by wealthy donors seeking to advance libertarian ideas. As Daniel T. Rodgers argues, this new ideological ground would subsume hitherto political debates over questions of power to decide the distribution of rights and duties to a context-free and ahistorical jargon of consensus. This jargon would be anchored by thin moral injunctions that implored individuals to support vaguely environmental abstractions such as holistic awareness, well-being, and, indeed, sustainability.[2]

Early on, the radicals linked objectively measurable issues such as pollution and nuclear fallout with a growing tendency amongst large sections of the population to reject mass-industrial employment, militarism, and mass-consumer culture as, indeed, one-dimensional, unfulfilling, and unsatisfying. For environmental historian Robert Gottlieb, radical Greens led a generational revolt against established values. This revolt had spread not only across the United States but across all of the postindustrial societies from the 1960s onwards. For Gottlieb as well as for historian Barbara Epstein, dispirited New Left activists who had come of age in the participatory-democratic, civil rights, and anti-war movements began to respond to what critical theory had elaborated as the pathologies of Enlightenment modernity by fashioning normative guidelines for a peaceable green-left movement. As the radical coalitions of the 1960s began to fall apart amidst state-facilitated and, to some extent, business-sponsored suppression of dissent, it was a new wave of ecological thinkers and activists who enunciated a highly appealing critical diagnosis of the times. Such a diagnosis attributed both social injustice and environmental pollution to the primacy of what the foundational text of critical theory, Horkheimer and Adorno's *Dialectic of Enlightenment*, had described as the reenchanting myth or alibi of Enlightenment, the ideal of progress towards freedom through the collectively organized rational mastery of nature.[3]

For Gottlieb, the syncretic relationship between remnants of the New Left and the nascent Green movement was extremely important. For Gottlieb, the broader left had long harbored radical social-reform movements that opposed concrete instances of human exploitation, including pollution hazards, as pathological consequences of modernization. Predominantly mobilizing in the densely populated industrial cities of the Northeast from the late nineteenth century onwards, radical social reformers had linked factory runoff, such as smog, sewage, chemicals, and manure, to issues within factories, such as occupational injuries, exposure to toxics and the low wages that accompanied unregulated sweatshop conditions. These radicals, on the left but for the most part working outside of the organized labor movement, had identified such pathologies with the prevailing political order. The response was not only to provide aid to those mostly poor and largely minority individuals who suffered the most but also to build new exemplary communities in which clean air and safe working conditions would prevail. In this view, it was only in the late 1960s, with the building of alliances between these older radical social reform movements and new social movements organizing around the critique of Enlightenment dualism, that these pathologies increasingly came to be gathered together under the abstract category of "the environment."[4]

As the symbol of widespread public opposition to despoliation of the environment, the first Earth Day in 1970 was in fact the summation of what for Gottlieb were "contradictory impulses." Green-left new social movement activists as well as people as diverse as Democratic Party congressman Gaylord Nelson, consumer rights advocate Ralph Nader, and direct mail marketing entrepreneur Stewart Brand supported Earth Day as an effort to criticize the system through peaceful mass mobilization. Meanwhile, "socially responsible" business leaders lent support in an effort to incorporate the conservative, respectable silent majority, and pro-market "corporate point of view into a newly defined environmental consensus." Even Republican president Richard Nixon publicly recognized that "the great question of the '70s is, shall we surrender to our surroundings or shall we make our peace with nature and begin to make reparations for the damage we have done to our air, to our land, and to our water." Indeed, the Nixon administration publicly extolled "environmental reform, including protection of public lands, sought to enlarge federal responsibilities, avowedly for the good of all Americans," and emphasized "the shared threat to the environment and the public welfare, and the national interest in addressing environmental degradation."[5]

This greening of America chimed well with broader public desires for an irenic, conciliatory, and citizen-centered politics that could replace the dysfunctional insider politics of Washington and the states that was prevalent in news media reports of the period. Support for the new environmental consensus crystallized as radicals continued to draw links between the modern left's collusion in the discriminatory implementation of the New Deal and Great Society programs and support for the ongoing conflict in Southeast Asia, and its unquestioning fealty to idealized images of the white male breadwinner, suburban lifestyles, and stifling cultural conformity. Although opposed to mainstream involvement, the radical Green movement's participation in Earth Day helped to prove to a divided nation that in fact everything is connected to everything else. By the early 1970s, radical Greens had successfully "broadened the definition of domination [and so freedom] to include human exploitation of nature, and . . . extended the notion of community to the entire ecosphere." Greens had successfully drawn into public view a new value system that privileged the interconnectedness of society with nature. The once definitive Enlightenment belief that humanity was rightful master of nature had been challenged. In its place was offered a value system that prioritized an altogether more amorphous belief that humanity should be understood as but one immersed participant within the whole ecosphere, as denizen of a fragile Blue Marble or resource-constrained Spaceship Earth.[6]

In this light, the radical green deep and social ecologists are important because they directly challenged both the pathologies of modernity and the modern left's longstanding focus on simply redistributing the spoils of industrial capitalism's exploitation of nature. Deep and social ecologists elaborated a powerful critique of the radical Marxian and reformist liberal-democratic left's commitments to the failed Enlightenment project of organizing rational mastery over nature to universalize freedom. For deep ecologists, the agents of the impending green revolution would not be Washington bureaucrats, corporate managers, and members of labor unions but those voluntarily "'dropping out," embracing nonconformist lifestyles, and "tuning in" to nature. For social ecologists, ecological science and awareness that alienating social hierarchy depended on a false dualist premise had distorted Enlightenment's true holistic potential. Aware of such distortions, social ecologists saw ethically motivated individuals as responsible for building self-governing and egalitarian local communities. By dissolving human-human domination, such eco-communities would foster a decentralized social ecology in which human-nature domination would also as a consequence cease.[8] Whether deep ecologists exercising "idiosyncratic self-realization,"[9] or social ecologists cognizant of "the organic consociation of human beings,"[10] these radical participants in Radkau's first great eco-boom of the 1970s sought in fact to disenchant the Enlightenment myth of progress through rational mastery.

Given this generic background, it is necessary to consider further the specific ontological commitments, epistemological presumptions, and ethico-moral claims elaborated by both deep and social ecologists. In addition to the similarities, two differences between deep and social ecology are important in this regard. The first is philosophical and relates to the terms on which each would elaborate a holistic ontological commitment aimed at displacing Enlightenment dualism. Deep ecologists defended ecocentrism, the view that prevailing values should be nature- and not human-centered. Wanting to "see through the erroneous and dangerous [dualist] illusions . . ." of Enlightenment, deep ecologists sought a holistic philosophy that

> goes beyond a limited piecemeal shallow approach to environmental problems and attempts to articulate a comprehensive religious and philosophical worldview. The foundations of deep ecology are the basic intuitions and experiencing of ourselves and Nature which comprise ecological consciousness.[11]

The holistic interconnectedness and interdependency of each individual with everything else is in the deep ecological view the natural state of consciousness that modernity dissolves and which we should strive for. In contrast, social ecologists opposed ecocentrism and avowed what Murray Bookchin called naturalism. Bookchin's naturalism, a weak version of anthropocentrism—the view that values humanity over the rest of nature—would support the evolutionary enlargement of humane, cooperative, and immersive social relations. This evolutionary enlargement would undermine the growth of pathological, utilitarian, and exploitative social relations. For Bookchin, "social ecology, in effect, is the concept of an ever-developing universe, indeed, a vast process of achieving *wholeness*," such that activist awareness of ecological holism sustains the egalitarian consociational ethos necessary to steer human history back into alignment with its true developmental potential.[12]

The second difference between deep and social ecology is ideological. Early on, deep ecologists such as Judi Bari had at times with great animus explicitly rejected all forms of leftism as iterations of a hubristic desire to dominate nature. In Bari's view, the emancipatory project to which radicals should dedicate themselves was one of planetary, ecospheric liberation. Deep ecology goes beyond what Arne Naess had described as the left's shallow concerns with human emancipation. Over ensuing decades, however, most deep ecologists would abandon Bari's factionalism and increasingly agree with Dave Foreman's view that deep ecology was indeed a contributor to the broader left's historical project of universalizing freedom as autonomy. In contrast, social ecologists always promoted their movement as the natural historical next step for the radical Enlightenment project. Indeed, Bookchin had in the early 1960s developed the case against the modern Left and Marxism in particular on grounds that the doctrine was blind to the planetary destruction wrought by mass-industrial society. For social ecologists, Enlightenment values had been perverted by those who stood to benefit from industrialization, be they bureaucratic managers, corporate capitalists, or the "yellow" unionized workers who uncritically supported the system. As did Horkheimer and Adorno, both factions of this nascent green-left modernist movement regarded dualistic Enlightenment aspirations to the rational mastery of nature as a delusion built on civilizational hubris. The deep and social ecologists sought a return to the humility that comes with an immersive holistic, and so ethical understanding of the world as a complex whole that escapes full human knowledge and agency.

Drawing such a philosophical basic intuition or concept from the holistic critique of Enlightenment dualism, deep and social ecologists prioritized a distinctive approach to politics and morality. Radical Green-leftists prioritized answering not a political but an ontological question: What should take precedence, the parts or the whole? The answer to the ontological question lay in the commitment to the holistic conception of ecological reality: everything is connected to everything else. In this answer, what matters most is simply the ecosphere (deep ecological ecocentrism) or in a more nuanced sense, the ecosphere's potential to sustain a harmonious society (social ecological naturalism). Either way, the holistic response to the ontological question in turn raises a particular epistemological-ethical question: What is the status of the agents that grant precedence to the whole? As will be shown, after answering the ontological question in favor of holism, radical Green-leftists' response to this epistemological question is moral volunteers whose holistic eco-awareness obligates them to act ethically. Such eco-awareness is regarded as something innate to truly free individuals (in deep ecological crypto-Darwinism) or as a moral persuasion that can be learned by anyone (the social ecological re-theorization of Marxian of false consciousness). That is, both privileged the autonomously arrived at needs, desires, and actions of morally choosing individuals and their ethical communities over the structural constraints of the political order under which such individuals and communities operate. This is an epistemological presumption that grants higher normative status to individuals as morally free agents than to the political structure of society that constrains or facilitates any given individual's freedom. Both deep and social ecologists envisioned the normative task of dissolving nature-culture dualism and replacing it with immersive holism. For both the deep and social ecologists, the task was to mobilize an ethos through which individuals and local communities could express natural as opposed to artificial processes and relations.

In the wake of the 1960s' so-called Age of Aquarius, radical green-left efforts to combine holism with ethico-moral voluntarism found receptive audiences across the United States and the West. Such holism had long been an influential part of countercultural responses to prevailing dualistic ideologies. In fact, some deep ecologists even drew on the premodern Abrahamic traditions of Kabbalah, Gnosticism, and Sufi to represent holism and conciliation of the self with the true cosmological order as alternatives to Enlightenment dualism.[13] However, perhaps the most effective tool for elaborating a green-left ethos from holism in the 1970s was that provided by Spinoza's Enlightenment counterculturalism. Spinoza had problematized nature-culture dualism by arguing that all is nature, including the cosmological order that monotheistic dualists attribute to the will of the deity. The implication was that individual moral choices to conform to the ethical dictates of nature (and so of the deity) sit under the dominion of the individual. Should an individual intuit the dictates of nature, the resulting moral choices are not in any way subject to regulation by external authority. Indeed, what ultimately matters are not the dictates of external authority but the inner worldly dictates of piety Freedom is to be found in personally understanding the necessities of nature. Interest in Spinoza's once subversive views had gained significant impetus amongst countercultural thinkers in tandem with the growth of modernity from the eighteenth and into the nineteenth centuries onwards. Spinozist holism experienced a second revival in the late 1960s and 1970s. As an aid to the disenchantment of the Enlightenment myth of progress through rational mastery over nature, Spinozism allowed deep and social ecologists to link the holistic critique of dualism to a moral voluntarism that rejects what critical theory had interpreted as the administered world.[14]

Filtered through the United States' own unique countercultural history, which extended from nineteenth century transcendentalism and anarchism to twentieth century pragmatism and agrarian populism, deep and social ecologists used spiritual connectedness or ecological science respectively to grant to the holistic ontological commitment a normative ethico-moral function. Alongside insights drawn from critical theory, some deep ecologists sought to link Spinozist tropes with Heideggerian philosophy, novel cybernetic and systems theories, process philosophy and pseudo-scientific theories such as the Gaia Hypothesis. Others embraced various holistic Eastern and Amerindian spiritualities as well as hip and beat philosophies. In similar yet more rigorously dialectical materialist terms, social ecology channeled the holistic critique of Enlightenment dualism into a complete anthropology designed to challenge critical theory itself. For Bookchin, such an anthropology would yield a more genuinely "critical . . . integrative and reconstructive science [because] in the final analysis, it is impossible to achieve a harmonization of man and nature without creating a human community that lives in a lasting balance with its natural environment."[15]

So understood, efforts to combine ontological holism with the epistemological prioritization of the moral volunteer led Green-leftists to understand and act within the society-nature relationship in a distinct and highly individualistic way. What I call the radical green-left synthesis represented this relationship as an unbroken continuum that extends through individuals across the planet and backwards and forwards in time. The moral choices of eco-aware individuals, by coming together to constitute ethical eco-communities, guarantee movement towards a harmonious community of communities on a planetary scale. Indeed, deep ecological "non-conformists" who chose to exercise "ecological consciousness" were understood as pioneers of a self-reproducing global community that worked in "ecological harmony or equilibrium." For Bookchin, the "basic communal unit of social life must itself become an eco-system—an ecocommunity . . . diversified, balanced and well-rounded

[and] on a social scale that always remains within the comprehension of a single human being."[16] Such positive beliefs—shared with other radical left movements at the time, notably, anarcho-syndicalists, autonomous Marxists, and Gramscians—implied that "intentional" moral choices, put into practice in local, autonomous communities, could and would "prefigure" a new society that conformed to holistic ecological principles.[17] For some deep ecologists, this meant that the ecosphere should be regarded as a whole "spiritual" nexus of being. For others, holism implied a "land ethic" that extends the ethical and moral "boundaries of the community to include soils, waters, plants, and animals; or, collectively, the land." For social ecologists such as Bookchin, "wholeness is, in fact, completeness. The dynamic stability of the whole derives from a visible level of completeness in human communities, as in climax ecosystems."[18]

The tendency for such a normatively charged combination of holism with voluntarism to obscure structure and institutions is therefore explicit in deep ecology. For deep ecologists, the job of attaining freedom is a personal project of achieving individual, and so societal, harmony within nature. Deep ecological holism called forth innate subjective awareness of immersion within a nature that modernity had corrupted. For deep ecologists, wherever exploitation is observed or experienced, it can be traced to the alienating effects of the illusory ontological commitment to dualism. Deep ecologists' resistance to mass employment and consumerism and their rejection of the conformity of the modern left is a rejection of modernity itself. The normative argument is that morally worthy actors choosing to replace paid work with self-help and commodities with handicrafts conform more closely to the true holistic ecological order of things. By dint of such eco-aware choices, deep ecologists will voluntarily place themselves and thus society in an ecocentric, right relationship within the ecosphere. Such rightness would emerge at the micropolitical scale of local communities that would instinctively replicate the deeper folkways of the whole ecosphere's community of life, or at the level of what Ulrich Beck would later define sociologically as the "subpolitical."[19] That is, eco-aware individuals would prefigure holistic ecological harmony. In these terms, deep ecologists laid blame for society's broaching of global limits to growth at the feet of Enlightenment's one-dimensional man, because

> ecological consciousness and deep ecology are in sharp contrast with the dominant [dualistic] worldview of technocratic-industrial societies which regard humans as isolated and

> fundamentally separate from the rest of Nature, as superior to, and in charge of, the rest of creation. But the view of humans as separate and superior to the rest of Nature is only part of larger cultural patterns. . . . Western culture has become increasingly obsessed with the idea of dominance: with dominance of humans over nonhuman Nature, masculine over the feminine, wealthy and powerful over the poor, with the dominance of West over non-Western cultures. Deep ecological consciousness allows us to see through these erroneous and dangerous illusions.[20]

For deep ecologists, to universalize conditions for freedom as autonomy, individuals needed to "reach inside" and find accordance with "the rest of creation."

The normatively charged analytic takes a more circuitous and arguably nuanced route in social ecology. The social ecologists drew on anarchist themes to urge free individuals—who recognize that modernity is founded on a falsehood that justifies artificial hierarchical relations—to establish autonomous, self-regulating organic eco-communities. Social ecology in this way renders problematic both the liberation of the self sought by deep ecologists and the mass or class liberation sought by the modern left.[21] To quote Bookchin at length,

> Having emerged out of a long process of evolutionary development in which [humans] were often subject to so-called "natural laws" in an evolutionary phase that we call "*first* nature," humans have created a cultural and social line of evolution of their own. This evolution is based on highly institutionalized societies that I have called "*second* nature." [T]he human species—itself a product of natural evolution—is no longer simply subject to "natural laws." Human beings can play an appallingly destructive role for non-human life-forms, or by the same token a profoundly constructive role. This is not preordained by "natural law."
>
> Our ecological dislocations have their primary source in [the] *social* dislocations [of second nature]. Granted, we need profound cultural changes and a new sensibility that will teach us to respect non-human life-forms; that will create new values in the production and consumption of goods; that

> will give rise to new life-fostering technologies rather than destructive ones; that will remove conflicts between human populations and the non-human world; and that will abet natural diversity and evolutionary development.
>
> Over the centuries . . . social conflicts have fostered the development of hierarchies and classes based on domination and exploitation in which the great majority of human beings have been as ruthlessly exploited as the natural world itself. Social ecology . . . reveals that . . . the very *idea* of dominating nature stems from the domination of human by human [in second nature]. This hierarchical mentality and system has been extended out from the social domination of people—particularly the young, women, people of color, and ues, males generally as workers and subjects—into the realm of non-human [first] nature.[22]

By rejecting the modern left's fetishization of state capture and instead relying on self-managing, indeed subpolitical, eco-communities as the route to universal freedom, social ecologists modify György Lukács's original formulation and seek to *align* second with first nature. Social ecology's eco-aware individuals are thus building a

> decentralized society, not only to establish a lasting basis for the harmonization of man and nature, but also to add new dimensions to the harmonization of man and man. . . . [This is because a] relatively self-sufficient community, visibly dependent on its environment for the means of life, would gain a new respect for the organic interrelationships that sustain it. . . . [It follows that] dissolving propertied society, class rule, centralization and the state [will] yield a sensitive development of human and natural diversity, falling together into a well balanced, harmonious whole.[23]

Social ecologists' path to liberation necessarily "involves above all a social process . . . the liberation of daily life profoundly concerned with lifestyle" and, as such, is blazed by moral volunteers who prefigure in the here and now the coming "destruction of power."[24]

Without doubt, the deep and social ecologists gave particularly acute expression to a significant and effective wave of public and intellectual protest against the prevailing ideology of their time. This was the general air of protest that Earth Day had harnessed so forcefully. As Marcuse argued, such protest was directed against a value system that, by artificially separating society from nature and setting the former over the latter, had breached just and tolerable limits. For the anthropologist Louis Dumont, "widespread protest against the disruption of natural equilibriums ha[d] for the first time in [modernity] placed a check on modern artificialism," a check that was in part informed by scientific findings that had caused many to question Enlightenment values. What sympathetic observers such as Marcuse and less sympathetic observers such as Dumont together noted was that the nascent movements did not only seek to challenge pollution and waste. The protest movement also confronted a rampant and often callous instrumentalism that, it was felt, sustained the military-industrial complex, the centralized, bureaucratically administered welfare state and the supine conformity that had until then provided the taken-for-granted recipe for the good life. Green-leftists in this sense helped to constitute genuinely *new* social movements that were rejecting the Enlightenment valuation of society over nature in favor of a holistic ecological valuation of society as but one participant in the ecospheric system.[25]

Deep and social ecologists thereby provided the groundwork for development of the broader ideological position that I define as *green-left modernism.* Such an ideological position is green insofar as focus is directed towards the discursive object of the environment. It is left insofar as focus is simultaneously directed towards the universalization of conditions for freedom as autonomy. And it is modernist insofar as agency is understood as being immersed within, rather than as seeking control over, the world external to the subject. The avowal by green-left modernists of the ontological commitment to holism, epistemological prioritization of agency over structure, and, in turn, normative justification for moral voluntarism over collective political action is accordingly of a kind with broader societal transformations taking shape in the late 1960s and into the 1970s. These transformations would over ensuing decades entail a wide-reaching shift in political culture from one that emphasized authority's capacity to facilitate structural constraints on agency to one emphasizing agency itself, and, therefore, a shift from an emphasis on political institutions to one in which emphasis would be placed on individual moral choices to construct an ethical home within the pathological system that was modernity. At issue therefore is the way that radicals at the forefront of Radkau's first great eco-boom effectively worked to dissolve the problematic that Williams characterizes as the "first political question" of modernity. At both the philosophical and ideological levels, this green-left *modernism* presumes that the commitment to ontological holism and concomitant discharge of moral obligations by individuals in ethical communities can and should dissolve the need for collective organization. That is, it fails to see the necessity for any resolution arrived at through politicized contestation with those who do not automatically accept the radical or indeed reformist projects' goals of universalizing freedom.

The Communitarian Turn

Throughout the 1970s, debates amongst radical green-left modernists deeply influenced and were in turn deeply influenced by broader intellectual debates. These debates were shaped by three interwoven problems. Each of the three problems concerned questions of how best to interpret the relationship between society and nature, and, in turn, the status of Enlightenment values. The first problem turned on assertions that the modern left had failed to challenge what critical theory saw as the pathologies of Enlightenment. The modern left had for too long championed nature-culture dualism, which in turn had led to a fateful embrace of rational, calculative egoism. As argued, it was this critique of the modern left that was central to the synthetic green-left alternatives offered by deep and social ecological modernism. The second problem centered on claims that Enlightenment dualism and its concomitant epistemological prioritization of structural constraint over individual agency were somehow inaccurate, wrong, or false, and should be rejected. Again, critiques of dualism and calculative egoism were central to both deep ecologists' calls for a new consciousness and social ecologists' calls for those seeking justice to build proximate eco-communities. The third problem centered on arguments that the only way forward for the left was to eschew political authority as artificial and so irredeemably corrupt and in its place to reconceive authority as a moral continuum, wherein obligations that fall directly upon ethical individuals acting in face-to-face communities fall equally upon states, their agents, and the institutions that they sponsor. Again, such a critique of dualistic distinctions between political and moral authority as artificial was central to deep and social ecology.

A coherent response to these three problems would emerge in the 1970s. This response materialized under the aegis of a wide-reaching shift on the left, away from modern emphases on the state and towards a revival of civic-republican emphases on citizenly self-help and the normative influence of traditional lifeways. The philosopher Charles Taylor offers a contemporaneous account of the temper of a time in which the radical communalist spirit that had in the 1960s attracted people to leftist and countercultural strongholds such as San Francisco's Haight-Ashbury and Manhattan's Greenwich Village began to reverberate back out across the United States and the West in general. As had Marcuse and Dumont, as well as avowed counterculturalists such as Charles A. Reich and Theodore Roszak, Taylor explicitly drew links between the failures of the modern left and the increasingly diffuse sense of cultural malaise that was fuelling widespread "ecological protest against the unreflecting growth of technological society." I concentrate on Taylor's and fellow communitarian revivalist Alasdair MacIntyre's work here for a few reasons. Both Taylor and MacIntyre had been central figures within the intellectual movements that had informed much New Left thought and practice since the early 1960s, both in North America and Western Europe. Taylor directly eschewed dualism and utilitarian calculating egoism and in its place offered an alternative holistic, "expressive individualism" that regarded people as immersed in, shaped by, and potentially able to shape their communities, and so their environments. He also readily incorporated ecological arguments into his political philosophy and activism. MacIntyre similarly advocated a turn away from the state and saw proximate communities bound by an ethics of virtue as the only viable response to the collapse of Enlightenment under the weight of its own contradictions. MacIntyre defended a post-Marxist account of community and work as the last bastions of timeless, nature-derived Aristotelian virtues and posited these as defenses against an alienated and totalizing modernity. Although MacIntyre all but ignored the philosophy of nature, the grounding of his ethical case in narrative and tradition, and his rejection of egoistic utilitarianism and liberalism, makes him relevant here. Moreover, a number of contemporary green scholars see MacIntyre's work as providing the foundation for a genuine ecological ethics.[26]

Just as radical greens had challenged the modern left, Taylor and MacIntyre challenged the dominance of state takeover philosophies in left thought and practice, both in Marxism and liberalism. For them, Marxism and liberalism had become indistinguishable from the increasingly authoritarian tendencies of rationalist utilitarianism. Along with other communitarian revivalists such as Michael Sandel and the relatively liberal-minded Michael Walzer, Taylor and MacIntyre challenged both as having evolved since the nineteenth century into justifications for the bureaucratic state and corporate capitalism.[27] In Taylor's view, people on the modern left had for too long advanced and bought into instrumental utilitarianism and liberal contractualism as the sole means for universalizing freedom as autonomy. It had become clear by the 1970s that the Enlightenment dualist presumptions informing both depended ontologically, epistemologically, and therefore ethically upon the flawed

> construal [of] the subject as ideally disengaged, that is, as free and rational to the extent that he has fully distinguished himself from the natural and social worlds, so that his identity is no longer to be defined in terms of [ontological continuity with] what lies outside him in these worlds. . . . [This is] the punctual view of the self, ideally ready as free and rational to treat these worlds—and even some of the features of his own character—instrumentally, as subject to change and reorganizing in order the better to secure the welfare of himself and others. . . . The social consequence [is the] atomistic construal of society as constituted by, or ultimately to be explained in terms of, individual purposes.[28]

For Taylor, it was high time that the intellectual and activist left began to question this adherence to instrumentalism and to embrace a literally countercultural alternative. This alternative was to be found on the streets, in the waves of rebellion sweeping the advanced industrial societies: the Anglosphere, Europe, and Japan. The culture of rebellion against existing values was giving voice to what Taylor saw as an "expressive" alternative to instrumentalism. Taylor ties this expressive individualism to nineteenth century Romanticism and to Hegel and contrasts it with both instrumental utilitarian and rights-based individualisms.[29] Where the instrumental calculating egoist had sought to measure, order, predict, and control nature in order to maximize happiness, the expressive, indeed modernist, individual seeks to be at home within the pathological condition that is modernity.

As had Marcuse and Dumont, Taylor noted that what was taking place was a political, cultural and ideological transformation. This transformation extended well beyond the academy and into society. Support for calculative utilitarian egoistic individualism was waning. Widely felt fatigue with the stalled progress of liberal rights had set in. A holistic expressive modernism was in the ascendant. Importantly for Taylor, the new social movements and especially radical ecologists were not to be understood as traditional romantics, yearning for a simpler past. Even though at the time some deep ecological primitivists and back-to-the-land eco-separatists did evoke such motifs.[30] The wider sensibility that the social, if not all deep, ecologists tapped into was forward looking and indeed expressive. For the communitarian Taylor, as well as radical Greens and leftists, and their sympathizers and those they hoped to mobilize and represent,

> Human life is seen as the external expression of a man's potential. This view, which we see developing in Rousseau, Herder, into Romanticism, Hegel, Marx, eventually infecting even the utilitarian tradition, through Mill, does not return to pre-modern models of man. The potential which a man expresses is very much his own; it develops out of him, and is not defined by some relation of harmony with a larger order. . . . Life is seen also under the categories of expression, as being a true embodiment of potential, or a distorted, compressed travesty of what men truly are. . . . By the same token, men's relation to the surrounding world is not simply seen as manipulative [but as] an ideal of communion between men, and between men and nature . . . Men sought to recover a relation with nature where they sense it, too, as expression—of spirit, life-force, or whatever—which their lives, as expression, are in tune.[31]

In Taylor's view, the difference between premodern models of man and the expressivism that he witnessed unfolding in the 1970s was that "on the earlier views man could only come to himself by finding his right relation to the larger order, whereas now men reach communion with nature by discovering what they really have it in them to express."[32] In these terms, however, Taylor's expressivism seems merely to invert the malignant Darwinian presumptions about human nature at the core of the utilitarian calculus. What expressive modernism expresses is the essence of human nature, conceived as harmonious communion between men, and between men and nature. Into the direct holistic relation of harmony with a larger order that defined the premodern model of man is inserted the inner worldly moral choice that Spinoza had awakened at the dawn of modernity. In this sense, what Adorno had problematically spurned as the mimetic yearning, and condemned as a "regression" is also applicable to Taylor.[33] Taylor's expressivism imitates and indeed expresses some idealized, benign cosmos. Expressivism is the form taken by mimetic yearning when in tune with scientific ecology and the existential experience of relative subjective ease within a cosmopolitan, multicultural world.

What thinkers such as Taylor and MacIntyre noticed and contributed to in the 1970s was a rethinking of the universalistic commitment to freedom. No longer regarded as the product of egoistic choices or state-sponsored rights, freedom was reconceived as the product of individual moral choices to act in local communities and so ethically prefigure humanity's rightful status within the encompassing order. In the radical green-left and the communitarian modernist views, radicals and even reformists needed to embrace, articulate, and elaborate a new politics and a new ontological commitment. Not only had the modern Left adopted practical reasoning based on the egoistic utilitarian calculus but also the dualistic Enlightenment ontology that stood behind the bureaucratic welfare state, sustained its limited rights, its consumerism, and its military-industrialism. Even though Marxists, Social Democrats, and Liberal Democrats disagreed on normative questions—on what constituted a just means for distributing the spoils of industrially organized mastery over nature—the modern Left as a whole had accepted its role as guarantor, through the state, of progress toward universal freedom through the rational mastery of nature.

What the radical Greens and communitarians elaborated so clearly was the suspicion that the modern left had wrongly presumed that the universe was comprised of objectively discernible natural forces. More so, they helped to show that the modern left had accepted the utilitarian and liberal premise that humans could justifiably deduce or reflect universal principles in order to measure, order, predict, and master nature for human ends. The modern Left had presumed that the state could do this without generating harmful consequences, or that in the case of harmful consequences, a correctly organized state could resolve these through the further application of such rationality. The capacities afforded to humanity by the evolutionary accident of greater sentience had falsely allowed some humans to imagine themselves as beings separate from and rightfully dominant over nature and others. As participants in the first great eco boom and coeval turn to community, deep and social ecologist and public intellectuals such as Taylor and MacIntyre helped to show a significant section of the population that the modern left's unquestioning commitment to Enlightenment dualism had in fact created and exacerbated gross pathologies. It had aided the creation of an order that was riding roughshod over possibilities for expressive self-realization and consociational communalism. And its dream of power had undermined possibilities for true freedom. Moreover, what the synthesis of green with left modernist and communitarian ideas had shown was that political institutions erected on the foundation of Enlightenment dualism could no longer be conceived of or treated as legitimate avenues for the pursuit of freedom. In response to this realization, radicals attempted to build local institutions, often justifying them in terms of their spontaneous character. The legitimacy of these truly subpolitical anti-institutions would not depend on some artificially constituted political authority. Rather, legitimacy would hinge on the degree to which the networks of local communities built by eco-aware moral volunteers promoted holistic symbiosis. These eco-communities would possess legitimate moral authority in direct proportion to the degree to which they and the moral volunteers who built them conformed to, reproduced, and ultimately expressed benign natural ecospheric processes.

Radicalizing Ontology

To reiterate, the holistic ontological commitment central to green-left modernism sustains a distinct epistemological presumption about the relationship between the whole and its constituent parts. It presumes that the ecosphere operates on principles that individuals, as parts within the whole, can and should voluntarily choose to express. The ecosphere is a unitary whole that, while not amenable to complete knowledge or total control, nevertheless demonstrates to activists and engaged citizens, to scientific ecologists and communitarian political theorists, what the right moral and political choices are. The whole is thus conceived as a seamless moral continuum. This is a positive version of the naturalistic Nietzschean negative insight that effective, heroic, moral choices lead to, or away from, virtuous status along the continuum. From within the green-left modernist viewpoint, lauded is the individual who chooses to recognize and act on the holistic ontological commitment. Such choices demonstrate worthiness within the ecospheric community of life. Authority is in this way reconceived as a function of eco-aware ethical individuals' moral choices. Communitarian efforts to instantiate social relations on ethical eco-aware principles appear just because they imitate truly natural processes. This is a conceptual leap from ontology through epistemology to ethics that renders moral authority coextensive with, and so an expression of, the benign natural order of things.

Arguably, for the first time since Enlightenment values contributed to the secularization of Judeo-Christian political culture several hundred years prior, a significant and effective faction of the radical left saw the material world, indeed the universe, as once more having "meaning in itself, as an ontological feature."[34] Morality and politics, as well as the authority to uphold them, were to express natural processes rather than demands born of human artifice. This is very important, because as Taylor himself recognized, to settle an ontological commitment or question first, prior to political engagement, is to implicitly shape an ethical position, and so options for practical advocacy. Such a move expressly lays out "the essential background of the view you advocate." Settling ontological questions involves determining "what you recognize as the factors you will invoke to account for social life . . . the terms you accept as ultimate in the order of explanation." Therefore,

> once you opt for holism, extremely important questions remain open on the level of advocacy; at the same time, your ontology structures the debate between alternatives, and forces you to face certain questions. Clarifying the ontological question restructures [and so restricts] the debate about advocacy.[35]

Settling on an ontological commitment to holism delimits the range or scope of possibilities for practical reasoning, that is, subjectively or collectively reasoning about the relationship between means and ends. A political actor who makes such an a priori commitment therefore restricts the range of available practical reasoning options. This is because "ontological accounts have the status of articulations of our moral instincts," and it is these instincts that shape what are acceptable as practical reasons for action.[36]

The use of an ontological commitment as the foundation for practical judgment or indeed, a political movement, raises the issue of the status granted such commitment by ideological partisans engaged in political contest. This issue becomes significant when considered from the analytic purview of the realist critique of ethics-first and moralizing approaches to politics that I derscribe at the beginning of chapter 1. In the realist lens, politics in modernity is not well practiced by partisans who understand engagement as a debate over principles, rather than as a contest to exercise power amongst parties that must at least be respected or even feared. In Geuss's oft-cited view, effective political engagement involves prioritizing action and the contexts of action, in particular the impacts of action on one's own and politically significant others' subsequent actions, as well as on institutions such as the state or markets. Indeed, Geuss is strongly critical of moral utopians, including the communitarians' left- and right-liberal opponents, John Rawls and Robert Nozick. Williams too criticizes Rawls and Nozick for avowing an interest in politics and power but in fact emphasizing the normative force of morality, while also taking to task the communitarians Taylor and MacIntyre for similarly regarding political obligations as coextensive with moral obligations.[37]

In this view, both contractualists and communitarian revivalists uphold the pretense that politics and morality are coextensive analytically. Rather, when understood against the backdrop of the West's peculiar historical indebtedness to Enlightenment values, it appears to be the case that a starkly real, if contingent and shifting, boundary exists between politics and morality. The placement of this boundary is determined, at any given moment and in relation to any given issue, by the balance of power that obtains amongst partisans engaged in a power struggle to influence authority in its rationalized guise. That is, the boundary is drawn not by facts, truth, the better argument, or indeed raw power, but by artificial institutions that have been organized to respond to the first political question. Partisans who see engagement in politics as a matter of getting clear on morality and ethics or indeed ontological commitment seem in this perspective to be one of two kinds. More often than not, such partisans are simply ineffective. Or they are very effective, as authoritarians who seek to reestablish a political system premised on raw power. In the latter case, such moralizing partisans seek to undermine political authority organized through détente.

Rather than focusing on ideas and beliefs about the true constitution of and rightful orientation to the world, effective political engagement—including political philosophizing and theorizing—might be better informed by interpreting practical reasons for action and the ideological justifications that actors offer. Of course, such practical reasoning and ideology may take a range of forms. Yet, as MacIntyre himself recognizes, in the historical context of the United States, the practical reasoning that informs politics for the most part has taken, and continues to take, one of three contrasting forms. These are the utilitarian calculation of costs versus benefits to oneself, supported by a malignant, usually Darwinian form of what Fred Block and Margaret Somers call the "naturalistic fallacy"; the view that authority is obligated to instantiate some inalienable, ultimate truth, supported by a conversely benevolent naturalistic fallacy; and tacit, often begrudging, support for some public welfare-oriented set of collective constraints on and stimuli to agency, which remains agnostic in relation to naturalistic claims.[38]

It should be clear at this point that the form of practical reasoning suffusing modernist green-leftism with ideological appeal is that which sees authority as somehow obligated to instantiate some inalienable truth. That is, authority is charged with the twin normative tasks of reconciling humanity with nature and human beings with each other, and so reconciling the holistic coherence of the natural order. However, as the realists point out, authority in modernity is unavoidably rationalized *political* authority, organized not to express nature or the views of any individual or collective in particular but to uphold some artificial version of political freedom. Political authority is organized to respond to Williams's first political question and in doing so, of course, to favor some over others. As the deep and social ecologists and the communitarians point out, such responses have on the whole supported forms of political freedom that privilege the domination by a minority of the majority of citizens, and the gross exploitation of nature. In the green-left modernist view, political authority appears irremediably artificial and alienating for these reasons. It follows that political authority must be regarded as an unjustified, indeed an un-natural, interference in the natural order of things. The predominance of political authority in fact undermines individuals' potential status as free, morally choosing contributors to a harmonious and so ethical community-nature symbiosis. MacIntyre, Taylor, and the radical Green-leftists, each in their own slightly different ways, prioritize individual moral obligations to recognize, embrace, and ethically live out the natural order in their local communities. Each takes as a normative orientation the world as it should be or would be under the tutelage of the morally worthy. With this move, political freedom and other political problems—such as the curtailment of primitive freedom and its stablemate, raw power, or the structured maldistribution of welfare, or institutional unevenness in upholding benefits and obligations—are recast as moral problems of a somehow artificially deformed order, of ecospheric disharmony, and, it follows importantly, as moral failures of the human will to make an ethical home to inhabit.

Such practical reasoning that proceeds from ontology to normativity stems from what Adorno derided as false nonidentity thinking. As he saw, efforts to draw normative justifications for action from ontological truth tend to mask impotence or totalitarian ambition. As Taylor himself saw, at issue is that settling the ontological commitment to holism and combining it with moral voluntarism delimits the advocacy options available. Beginning from a benign or malignant commitment, an actor might virtuously yearn for some symbiotic eco-truth or for success in the Darwinian struggle. By answering the ontological question first, the green-left modernists' preference for ecospheric harmony, and so sustainability, and the Invisible Hand utilitarians' preference for maximal happiness and competitive success both restructure out of the realm of possibility that third historically available form of practical reasoning. This is the view that relegates settling ontological questions to second place in relation to collective agency oriented toward coercing rationalized authority to uphold some kind of contingent constraint on or facilitation of individual and collective interests.

In this view, the status granted individual agency in relation to the whole structure is reversed. Institutional arrangements take priority along with the requirement of obtaining some kind of best-worst settlement. Such constraint appears from the perspectives of practical reasoning guided by inalienable right or egoistic good as an artificial construction of political authority, and so unjustified. These absolute perspectives occlude possibilities for participating in settlement under the aegis of Williams's first political question on grounds that ethical, moral, or indeed ontological questions remain unanswered. They fault political authority for failing to establish an order that accords with truth. The idea that, in acting to serve its members' interests, a party might influence a political authority despite being unable to control it is pushed off the table. Most importantly, the "ontological" perspectives occlude possibilities for recognizing that authority is facilitated and constrained by political structures—institutions, constitutions, laws, regulations, conventions, and so on—that, while never impartial, are amoral insofar as they depend on the exercise of collective power to sustain them. Rationalized political authority oriented to some contestable definition of the public welfare, the best-worst option, cannot be peaceably established as the coextension of some Gaia principle, scientific fact, pure inalienable right, life-world ethos, or in terms of the pressures of natural selection.

When employed as a normative justification, holistic ecology obscures recognition that artificially constituted institutions—which apply some kind of power-dependent distinction between voluntary moral (spiritual, ecological, communal, virtuous, right, or good) obligations and binding (indeed, real deontological) political obligations—are legitimate foci for advocacy and action. Ontological holism restricts possibilities for acting collectively to support deontological *institutions*, as opposed to deontological conceptions of justice or rights, because it does not permit the political system to be viewed as, at least for the foreseeable future, a legitimately amoral, artificial authority. Rather, authority either appears to be moral or immoral. Dissolved are issues that the modern left had for better and worse addressed as political problems resolvable by institutions that upheld or dissolved political freedom in proportion to pressure applied. For example, the modern left's "compromise" with big business and big government had, where possible, ensured that the ends sought by certain institutions—such as labor-regulatory, food and drug safety, educational, and welfare bodies—were operationally partitioned from and could serve to constrain market competition. The modern left had parsed action in dualist terms, as public and political or private and moral, as citizen- or market-oriented. In that perspective, moral obligations are voluntary, with noncompliance inviting ostracism or exclusion, while political obligations are mandatory and invite what Weber called legitimate violence. So understood, holism confines advocacy to terms that imply the benign goods that ethical actors choose to enact can also potentially be exhaustively embodied by a large, amoral political structure, or that such a structure can be made to operate on principles that are coextensive with those held by ethical actors and which obtain in proximate communities.

Be it eco-communal and benign or Darwinian and malignant, the derivation of norms from an ontological commitment to holism makes it difficult to recognize an artificial political order that functions, or fails to function, due to contest over the definition of political freedom's boundaries. Importantly, this view undermines possibilities for understanding political authority as something that may legitimately stand as a potential barrier to raw power. Whether emanating from the Left or the Right, challenges to authority take on a utopian cast and sideline the more realistic view that such challenges contribute to a permanently unstable détente. From the 1970s onwards, much radical green-left scholarship and activism would reject engagement with political authority in favor of moral engagement oriented to refounding authority on truth. Opportunities for those seeking to obtain a share of what welfare the United States produced—sustainably or not—would be subsumed by intentional prefigurative agency aimed at creating a future-possible moral world "in the here and now."

As Luke argued in the mid-1970s, even the most radical green ideas risked being drafted into service as moralistic efforts to depoliticize issues such as pollution, climate change, or quality of life. For him, green-left activists risked involving themselves in a regime of "artificial negativity," that which Adorno saw in a philosophical lens as the false nonidentity thinking or positive negation that expresses rather than critiques raw power. The dawning of Radkau's age of ecology could, in Luke's view, offer only a soothing and appeasing narrative that fulfilled the bourgeois romantic, subjective ideal of a permanent escape from the antagonistic, deceit-ridden, and artificial world of politics. The spread of eco-awareness could offer the moral high ground not only to radical activists scarred by the collapse of the New Left, the stalling of an effective civil rights and peace movement, and dissolution of the urban counterculture, but also, unavoidably, to frustrated suburbanites and a new class of technically adept participants in the emerging infotech and service economy. From this ground, it would be a short path to an environmentally sustainable universe of ethically acceptable consumption and production. For radicals, the abandonment of politics and flight into intentional, prefigurative subpolitical communities would lead to a benign utopianism of communes such as Arcosanti, a proliferation of disconnected rural "hippie cults," the bohemian chic of ecotaging *Earth First!*, and its mannerist mirror image, the *Whole Earth Catalog*.[39]

Beginning in the 1970s, many radicals and, eventually, reformists alike would deride the idea that political power and so the achievement of green-left ends depend on the institutionally sponsored redistribution of pieces of the pie. Such ends were derided as symptomatic of scholarly and practical seduction by Enlightenment's dualist delusions and the productivist ideology. Obscuring reference to historical context or political structure in favor of the essential truths of ecospheric harmony and ethical eco-awareness, green-left modernism would orient its advocates to the task of correcting an immoral order. Individuals who make the moral choices necessary to organize social relations along ethical lines would achieve this moral corrective in local eco-communities. Such eco-communities would prefigure, in the present, the order that should obtain in a truly holistic planetary community. Radical green-left modernism would orient generations away from political contest and towards efforts to live rightly amidst what Adorno had argued was a "wrong life" that did not admit of living rightly. At the time, perhaps, the pathological consequences of the modern left's willingness to suborn broad concerns with universalizing freedom to the narrowness of the utilitarian calculus were all too evident for things to be otherwise.

What I am proposing here is that as an ideology, green-left modernism propounds a line of reasoning that closely resembles Williams's concept of a structural moralizing philosophy, theory, or justification for action. To reiterate, a justification for action on "the model of structural moralizing lays down moral conditions of co-existence under power, conditions under which power can be justly exercised," such that "morality offers constraints on what politics can rightfully do."[40] Practical reasoning on the model of structural moralizing defines the ends of action in terms of a required structure. The objective of agency informed by structural moralizing is to specify the right and to hold it up as the definitive constraint that should apply to authority. Some essential, inalienable truth (in the case at hand, benign ontological holism coupled with moral voluntarism) is taken as the basis for remaking, indeed, purifying authority. Interesting is that not only green-left modernism but also Taylor's and MacIntyre's communitarianism, can be conceived as examples of structural moralizing. The communitarians' relativizing of authority to tradition and local lifeways stand in contrast to liberal and libertarian defenses of pure inalienable right. Yet, both treat political authority as the linear extension of moral authority. In the former, the sovereign republican citizen and, in the latter, the counterfactual truth adhered to by partisans of the inalienable right to fair justice or private property specify the limits of permissible conduct under power.[41]

Interpreted through the lens provided by Adorno, a structural moralizing ideology or philosophy seems premised on what he derided as mimetic yearning. What is yearned for is, in effect, a world that once more has or gives-of meaning in itself, as an ontological feature. In the case of Green-leftists and communitarians, such yearning is for a world lost or destroyed by the rational egoism that proliferated under the truncated Enlightenment commitment to dualism. In the case of liberals and libertarians, the yearning is for a world that could and should be this way but is not, due to the malign effects on properly rationalized authority of radicals and romantics. From an Adornian perspective, the problem with such moralistic yearning is that in both guises its advocates must "dream of power" to implement their vision of justice. It is in just this sense that for Radkau, "the global rhetoric that characterized the 'ecological revolution' from the first great eco-boom betrays a dream of power beneath the surface."[42] Once the object of critique congeals—which the discourse or constellation of "the environment" did in the 1970s—negative efforts to confront domination risk becoming positive endorsements of albeit a newly minted essential truth. Adorno's "speculative moment" of truthfulness oriented by the experience of unfreedom is lost. The negative commitment to truthfulness in the face of domination risks being replaced by a positive commitment to truth as essence. Critique risks being assimilated to power. Once the preserve of those suffering, assimilated critique tends to simplification, rationalization, and, ultimately, recuperation in the hands of those empowered sufficiently to define the positive task of making things better.

Closing the Circle

Illustratively, from 1970 onward, what President Nixon had hailed as "the great question of the '70s" did indeed prompt a host of positive responses to the newfound awareness that humanity was but an immersed participant in the ecosphere. Two important positive responses that stand beyond the remit of green-left modernism and communitarianism were elaborated by eco-authoritarians and mainstream professional Greens, many of whom were former radicals who, finding eco-communalism ineffective, exhausting, or both, sought to "become part of the solution" being developed by new green institutions such as the Environmental Protection Agency.[43] While most deep and all social ecologists, and communitarians such as Taylor, had regarded the dictates of ecological science as being compatible with democratic government, eco-authoritarians and some deep ecologists rejected this possibility. That said, the misanthropic fringe of deep ecology amounted to a few dozen activists working in the Pacific West and Northwest. However, a range of classically trained economists and scientists such as Garrett Hardin, William Ophuls, and Paul Ehrlich authored best-selling eco-authoritarian responses to the problem of the environment. From a misanthropic premise, eco-authoritarians explicitly sought to override political authority by propounding a series of sometimes highly coercive policy prescriptions. These included the imposition of foreign aid sanctions on Southern nations refusing to implement population controls and the reduction or abolition of social welfare payments to the poor within the United States. Such prescriptions were based on principles said to be acceptable to any rational observer of the unfolding global tragedy of the commons.[44]

Eco-authoritarianism is in these terms most often thought of as an instance of consequentialist ends-oriented reasoning and opposed to utilitarian means-oriented reasoning. Eco-authoritarians combine a naive Hobbesian conception of malignant human nature with a "moral line in which stark and mutually exclusive choices exist based upon either avoiding the ecological crisis through authoritarian measures, or suffering it."[45] The point is, however, that only those capable of making the hard choices should get to do so. With this move, eco-authoritarianism morphs from a consequentalist to a utilitarian justification for authority. The good choices are those that are right for the people most likely to survive and so deservingly to prosper in the malignantly conceived state of nature. Such survivors constitute and support what eco-authoritarian William Ophuls saw as a "competent aristocracy." The aristocratic damp squib of "the poor will always be with us" is replaced by an activist attitude. This attitude regards the poor as justly being coerced into having fewer children or none at all, for example. Such an aristocracy positions itself as the deserving beneficiary of authoritarian choices. The utilitarian goal of achieving the greatest good for the greatest number, which Hardin had derided as a logical impossibility, is made possible by specifying in advance who deserves to be counted as making up the greatest number of people rightfully to benefit from the good use of limited resources.

Mainstream professional Greens would offer a more politically palatable positive response to the problem of the environment, and did gain considerable influence in Washington policy circles. Mainstream Greens would move inside the beltway in the wake of Nixon's reforms in the 1970s in order to appeal to the regulatory state as facilitator of a new green utilitarian calculus. Avoiding political contest, and disavowing radical small is beautiful eco-communitarianism, the mainstream professional Green movement more or less followed the path cut by the modern Left, and readily embraced practical reasoning on the model of enactment moralizing. To reiterate, a justification for how to get from means to ends on the model of enactment moralizing involves conceiving of authority as the instrument of the moral. As such, enactment moralizing greatly appeals to those in power who would seek to right the affairs of a society and have the capacity to do so. For Williams, "something always implicit in the enactment model [is] the panoptical view: the theory's perspective on society is that of surveying it to see how it may be made better."[46] For the enactment philosopher or ideologist, all actions that a freely choosing individual makes are subject to panoptical evaluation as contributions to the good society. This approach favors a prepolitical conception of the good and, from that conception, formulates the rights that a rational authority should make available to individuals. In Adornian terms, an enactment moralizing approach to politics entails the subsumption of all to the concept; enactment moralizing is identity thinking in action.

The problem confronted by mainstream Greens is that prepolitically defining the good narrowly, as the rational desire to maximize (green) goods and avoid (green) bads, assumes away possibilities that an environmentally rational authority may warrant critique by those disadvantaged in some way. Mainstream Greens' embrace of the regulatory state and consumer choice thus frustrates possibilities for organizing collective power sufficient to participate in the legitimation process underpinning political authority's capacity to respond to the first political question. Moving inside the Beltway, mainstream Greens sought ample authority to promulgate rules ensuring that individuals conform to a rationalized set of ecospheric demands. Meanwhile, the capacity of the broader Green movement was transformed. No longer potentially a mass movement, the Green movement fragmented into lobbies against pollution, for wild nature, against toxics, and for renewable energy, for example. More importantly, radical green-left modernist concerns were forced to the margins as off-the-wall demands, unsuited to rational, consensus-seeking debate.

While practical advocacy on the model of structural moralizing appears simply benignly apolitical, advocacy on the model of enactment moralizing is in these respects anti-political. In contrast to structural moralizing, which attempts to remake authority in the image of the worthy ethical agent, who by dint of free moral choice holds privileged access to inalienable truth, enactment moralizing seeks to re-form the individual in the image of a rational authority designed by the worthy. Whereas for the structural moralizer, ethical individuals or communities express some essential truth that authority should rightfully embody, for the enactment moralizer, authority is taken to express some unquestioned, essential good from which the right is then derived. In either case, political authority is envisioned as instantiating moral assertions concerning the ultimate good or inalienable right. In the structural view, the entire social structure should exhaustively embody the essential truth on which the most moral of actors choose to act. In the enactment view, the worthy should strive to make such structures operate on principles that delimit the dispositions of all actors.[47] Whereas a structural moralizing philosophy or ideology yearns for mimetic regression, enactment moralizing also entails a yearning. In the case of the latter, the yearning is for panoptical omnipotence, for the power sufficient to rationalize the political and social order in this way. In either case, such yearning seeks a universe that once more has meaning in itself, as an ontological feature. For the structural moralizer, a truncated notion of Enlightenment rationality is held responsible for having turned an Eden into a Hell for most people. Conversely, for the enactment moralizer, truncated Enlightenment rationality is lauded for turning a hell into an Eden for the deserving few.

My aim in this chapter has not been to fault radical Green-leftists and communitarians who sought to create a better world in the late 1960s and 1970s. Nor has it been to speculate on how the world should have been understood at that time. Rather, it is to emphasize how particular agents, mobilized by distinct experiences, in an identifiable historical context, helped to map a new political terrain. Advocates of green-left modernism interpreted the holistic critique of Enlightenment dualism for a receptive audience. In doing so, thinkers such as Naess, Bari, Foreman, Devall, Sessions, Bookchin, Taylor, and MacIntyre offered a coherent explanation for the societal pathologies that many were experiencing, not only as individual domination and environmental exploitation but also as self-inflicted psychic repression. Green-left modernism in this sense tapped into and echoed a major transformation. As recent work in political and cultural history shows, this transformation would eventually carry citizens away from what would come to be derided as reliance on the artificial support of the state and toward what would come to be lauded as self-responsibility in the face of adversity. Perhaps hyperbolic, it seems fair to say that this shift, in what Rodgers refers to as "the terrain of common sense," led an entire nation to gradually wish away the presence of impersonal structural constraints on agency and support for solidary organizations oriented to confronting political authority. Ultimately, the idea that political freedom defines a citizen's autonomy in relation to political authority would be displaced by the primitive conception of freedom. Freedom would once more come to be seen as the autonomy of the private person to make what is finally the moral choice to act in accord with natural forces.[48]

Against the backdrop of this transformation of a modern into modernist political culture, efforts to synthesize radical green with leftist and especially communitarian thought and practice would take a parallel turn. Greens and leftists would turn away from concerns with political authority, legitimacy, and contest and towards ethical concerns with morality, with individual intentions and prefigurative agency aimed at propagating social harmony and ecological equilibrium. Amidst the new contextless political culture, the once radical green-left modernist idea that an ahistorical and essentialist ethics should be combined with agentless and abstract universalist moral injunctions to uphold ecospheric harmony, equilibrium, and, eventually, sustainability would prove highly appealing to moderate center-leftists and compassionate conservatives. In seeking to "close the circle"[49] that Enlightenment had broken, Green-leftists and their neocommunitarian revivalist supporters would bring meaning back to the radical project. Ironically, however, the early iterations of green-left modernism had in the 1970s taken what is effectively a practical deontological position in relation to what was cast as illegitimate authority. The normative core of the ideology was that the inalienable truth which holistic ecology and moral voluntarism express would eventually undermine artificial political authority and replace it with a new sustainable and harmonious order. What could not be fully foreseen at the time was that, if holism and the injunction to support moral voluntarism were normalized, it would also need to be accepted that the kinds of external deontological critique that radical Green-leftists and communitarians subjected political authority to in the 1970s would be restructured out of the realm of possibility. Should the green-left modernist critique be fully realized, there would no longer be any outside, because "the circle" is closed—everything is connected to everything else. Put rather bluntly, and as Enzensberger, Radkau, and Luke foresaw, should science-derived holistic ecology and moral voluntarism come to define politics, there is very little that a green-left could do to challenge those powerful and influential actors who would justify their actions as being in harmony with or supportive of natural forces. In a world in which everything is connected to everything else, advocacy for institutions that would stand outside of and regulate social relations from a practical deontological position, and as such be rooted in interests different to those holding positions of power, would become very difficult.

Chapter 3

From Enlightenment Hubris to Neo-Enlightenment Humility

Legitimation Crisis

Whether interpreted by orthodox economists as an economic crisis provoked by deindustrialization, by heterodox political economists as a fiscal crisis prompted by government spending grossly exceeding taxation revenues, by international relations scholars as an energy crisis prompted by the sudden imposition of supply restrictions by the Organization of the Petroleum Exporting Countries, by conservatives as a crisis of democratic authority, or by neo-Marxist critical theorists Jürgen Habermas and Claus Offe as a legitimation crisis of "the liberal-capitalist social formation," the disruptions of the mid-1970s were both acute and of significant historical importance.[1]

In fact, the crisis of the 1970s was multidimensional. Both physical scientific and social scientific knowledge had each lately incorporated the ecological realization that the Enlightenment-sanctioned understanding of society as dualistically separated from nature could no longer be taken as given. A dissent-driven shift towards a post-positivist research paradigm was underway across the physical and social sciences. Advocates of a "new political science," for example, argued that

> One way to describe the crisis of legitimacy is to say that the basic features and tendencies of modernity have produced a situation in which the established processes and formal

> structures of control are at war with the conditions necessary for authority [based on dualistic, rational truth claims].[2]

As shown in chapter 2, informed by this nascent paradigm shift in the sciences, green-left modernist and reformist mainstream professional Green groups had gained significant levels of public support across the 1970s. Radicals and mainstream Greens had mobilized the citizenry sufficiently to convince Republican president Nixon of the need to act on the environmental problem, resulting in regulatory and legislative action on pollution control, limiting the use of toxics in manufacturing and food processing and increasing energy efficiency.

In another dimension of the crisis, many new social movement activists sought to replace mass-bureaucratic Taylorist business management models with various novel schemes. These extended from radical demands for full employee ownership of firms and worker self-management to flextime regimes and neo-Fordist shop-floor participation and reskill and retrain programs.[3] A further dimension of the crisis included a diluted subpolitical communitarianism that, taken up by large numbers of people seeking spiritual fulfillment, was used to justify a Marcusian Great Refusal of mass employment and consumption, militarism and chauvinism, racism and bigotry. Such new age movements rejected these variously as sources of spiritual dissonance and self-repression, and as being out of synch with natural processes and natural human desires for cooperative community-based lifestyles.[4] The times had indeed changed, and modernism had spread from its redoubts in universities, urban bohemia, the demimonde, and the art world to the suburbs and a generation that had experienced hitherto unprecedented access to higher education and high levels of disposable income, and who were, for want of better words, bored rigid.[5] Against this backdrop, and even while enacting environmental, consumer, and industrial reforms, both Republican and Democratic parties began to abandon welfare state and redistributive politics.

Witnessing this broad-based political cultural irruption, Habermas sought not only to analyze the "legitimation crisis" but also to use his interpretation as the basis for a renewal of critical theory. Habermas therefore avoided both Horkheimer and Adorno's pessimism and Marcuse's utopian optimism. He offered an alternative to their interpretation of the role played by Enlightenment values in producing the pathologies of modernity. For him, the combined demands of a restive public—enunciated not least by Greens and the modernist Left, but also by the civil rights, feminist, new labor, and various countercultural movements—had "overburdened" the truncated Enlightenment commitment to instrumental reason. The rationalized notion of efficiency that had sustained centralized, bureaucratic government and the increasingly multinational business corporation could no longer sustain the myth of progress through rational mastery. Such overburdening was the direct consequence of demands not only for a political economy that respected ecospheric limits to growth but also for increased government spending on public goods, including new institutions to promote and uphold civil, women's, and minority rights and expand access to higher education, healthcare, and welfare. As Habermas noted at the time and others have argued since, a new political culture had emerged. Habermas argued that,

> in the last ten to twenty years, conflicts have developed in advanced Western societies that, in many respects, deviate from the welfare-state pattern of institutionalized conflict over distribution. These new conflicts no longer arise in areas of material reproduction; they are no longer channeled through parties and organizations; and they can no longer be alleviated by compensations that conform to the system. Rather, the new conflicts arise in areas of cultural reproduction, social integration, and socialization. They are manifested in sub-institutional, extraparliamentary forms of protest. The underlying deficits [of the system] reflect a reification of communicative spheres of action; the media of money and power are not sufficient to circumvent this reification.[6]

The new social movements merely gave voice to widespread beliefs that life should be meaningful, authentic, and grounded in ethical self-awareness, while conducted in harmony within nature.[7] A significant and effective sector of the population in the United States and across the advanced industrial societies had embraced political cultural modernism. The legitimacy of the compromise between big government, big business, and big labor had been put to the test and found wanting. In Nancy Fraser's recent exegesis of Habermas's argument, by the mid-1970s citizens "had been primed to think that their states had the capacity and the responsibility to address . . . problems" that they held to be deserving of institutionalization. "Thus, they were disposed to hold [the state] accountable for . . ." a dysfunctional system that was "failing to deliver."[8]

As did Horkheimer and Adorno, Habermas had taken issue with the modernist new social movements.[9] He argued that the effort to base a political project on values "freed from immediate economic compulsion" and no longer bound by "the traditions of bourgeois morality" had indeed extended to breaking point the repertoire of social problems to which the state could respond without alienating capital. What the new social, green-left modernist and communitarian movements specifically demanded was control from below of an environmentally sustainable, community-based, classless, raceless, genderless, and so tension-free poststate authority. Those holding power at the core of the state-market nexus could not fulfill such demands without dissolving that nexus. In Habermas's opinion at the time, the new social movements rightfully sought such dissolution. But, in his view, the new movements were insufficiently powerful to gain purchase on political authority. This is because politicians and corporate owners and managers had only ever taken the modern Left seriously insofar as it could undermine state sponsorship of capital by provoking the withdrawal of labor power. Since the 1940s, the state and capital had cooperated to buy themselves out of legitimation problems through economic redistributive fiscal policy and social welfare programs, and, under pressure, haltingly extending access to such programs to restive minorities.[10]

Habermas therefore argued that the legitimation crisis of the 1970s was not about the state's incapacity or lack of willingness to respond to green or left modernist political, economic, or cultural demands. The crisis did not so much mark a point at which government recognized that it could no longer afford to extend redistributive and rights-defending programs to ever wider constituencies: local and especially urban communities, women, African Americans and Latinos, and youth, in the form of educational subsidies or unemployment and training programs. Rather, legitimation crisis marks the point at which government and business realized that they no longer needed to respond, as such, to secure mass loyalty, to secure legitimacy from organized blocs of citizens.[11] For Habermas, the crisis was about competing class interests and control or, abstractly, steering of the state. In the terms developed in this book, the crisis revealed to state and market interests the fact that radical and reformist movements no longer exerted power significant enough to shape the responses of political authority to the first political question. Such competition had been mediated by elected officials, who had for several decades presumed the need to secure the mass loyalty of one class while advantaging another, but who were fast realizing that they had fewer and fewer reasons for continuing to do so. Legitimation crisis thus marks the simultaneous decline of the modern Left as a political force and the recognition of this decline by politicians and corporate owners and managers.[12] As more recent work in political history shows, it was at this time that emboldened business leaders began to reevaluate their support for government as broker of compromise with citizens seeking to expand the contours of political freedom through collective action. And it was in the 1970s that the business community initiated a series of privately sponsored public relations exercises and scholarly efforts designed to call into question the progressive taxation and regulatory regime that upheld both the redistributive welfare state and the market regulatory apparatus.[13]

As Habermas had shown, since the advent of modernity, elected officials had been compelled by the need to advantage the owners and managers of capitalist firms. The interests of owners and managers lay in "capital accumulation, economic growth . . . institutionalized in an unplanned, nature-like way."[14] However, the "unplanned, nature-like" operation of capitalism depended upon the regulative capacity of the state. The aim had been to secure the mass loyalty of a unitary working class that, it was thought, could withhold labor and so undermine profit accumulation or, worse, embrace some populist ideology such as communism. As deindustrialization and the concomitant process of industrial downsizing gathered steam alongside the expansion of the postindustrial service sector, the modern Left's redistributive demands—once posing a credible threat and appeased through Keynesian redistributive borrow-tax-and-spend policies—were cast into political irrelevance. Simultaneously, it was recognized that the state could meet many new social movement demands without recourse to the redistributive policies that now appeared as a threat to capital accumulation rather than as a defensive strategy.[15]

Habermas's analysis of legitimation crisis informs what he would later elaborate as his theory of communicative action and subsequent discourse theory of law and democracy. Habermas developed his distinctive approach to critical theory in response to Horkheimer, Adorno, and Marcuse, radical phenomenologist Cornelius Castoriadis as well as to those he regarded as revivalists of the Nietzschean right, Luhmann, Foucault, Georges Bataille, and Jacques Derrida. For Habermas, these thinkers' post-structuralism or postmodernism was pushing critique and the radical Enlightenment tradition in general into the blind alley of a one-sided and totalizing subjective modernism. Such modernism was for him irrational, because derived from a precommunicative, presocial, and so subjective intuition-based "uninhibited skepticism" or, in the case of Foucault, from a contradictory Heideggerian anti-humanist yet subject-centered and totalizing reason. Habermas's alternative avoids what he calls the Nietzschean aporia altogether, including Adorno's efforts to develop a philosophical grounding for Left modernism. Instead, Habermas grounds critical theory's critique of domination as unfreedom in post-Kantian terms, as a critique of Enlightenment reason's failed promise.[16]

His account is therefore deliberately ambivalent about that which Horkheimer and Adorno had described as the Enlightenment re-enchanting myth of progress through rational mastery. On the one hand, the myth sustains the systematization of conditions in which domination is reproduced. Like Horkheimer and Adorno, Habermas finds that the artificial, rationalized authority of the modern, capitalistic liberal-democratic state is organized to defend private choices as expressions of the public interest in freedom, which in the process rationalizes freedom itself as the instrumental calculation of private costs and benefits. Yet, for Habermas, this same myth and the rationalization it produces also sustains the aspiration to freedom that Marx has conceptualized as species-being. Yet, Habermas goes on to reconceptualize species-being in terms of a communicative rationality that stands opposed to the instrumental rationality that is facilitated by the system. Communicative rationality persists in those areas of social life, the lifeworld, which at a given historical moment stand beyond systematization. In Habermas's theory, the noncoercive social relations said to obtain in the lifeworld therefore provide a normative—counterfactual and for this reason, potentially counter-systemic—template for fulfillment of the Kantian ideal of full autonomy or enlightened self-direction.[17]

Habermas's project depends on a binary analytic distinction between the interests located in the systemic world of work, of production, or instrumental action oriented to efficiency and those located in the life-world, wherein noncoercive communicative action oriented to truth obtains. Environmental philosopher Steven Vogel helpfully summarizes Habermas's argument:

> The institutional structure [of modernity, paradoxically] finds its normative basis in communicative processes; in premodern societies these are the traditions, myths, religious worldviews

> that serve to legitimate the form of social organization, while in modern society they begin to take on a more "rationalized" form, involving the notion of justification in terms of ideals of fairness or universality or human rights. In either case, legitimation of an element of the structure involves the application of "consensual norms" in which it is the normative rightness of the particular element that is put in question and in which it is the potential agreement of all affected parties that serves as the criterion of such rightness. . . . [Meanwhile] subsystems of purposive-rational action . . . represent spheres of activity in which the logic of work predominates. Here normative rightness is not an appropriate criterion but rather whether the proposed course of action is "successful."[18]

For Habermas, the trouble with modernity is that the instrumental subsystems—the techno-scientific, capitalist economic, and state bureaucratic—tend towards colonization of the lifeworld. Historically, this expansion occurs both objectively or collectively, through the subsumption of premodern social forms to what would come to be known as globalization, and, individually or subjectively, as careerism, consumerism, and fashion overtake pride-in-craft, durability, and ceremonial display. This colonization of the lifeworld by system-defined interests subjugates the communicatively rational interest in developing consensual norms native to the lifeworld. That is, system imperatives subjugate those consensual norms that emerge out of mutual understanding and express the interest in freedom as nondomination. The system imperatives that drive modernity are "no longer subordinated to an overall structure guided by communicative norms. The key to Habermas's reconsideration of Horkheimer and Adorno's early critical theory lay in his argument that the traces of true emancipation from domination, and so the potential for freedom's further development, are to be found in the very institutions of rationalized authority: in egalitarian rights, observance of the rule of law, and constitutional constraints on those holding power in government. These institutions represent the traces of communicatively rational lifeworld principles of egalitarianism, procedural fairness, and formal universality.[19]

As had Horkheimer and Adorno, what Habermas would retrospectively notice about the legitimation crisis was that it was a mistake to presume that the modern Left had uniformly been acting in the generalizable interest of all individuals. He concurred that Communists,

Social Democrats, and Liberal Democrats, as well as much of the organized labor movement, had bought into the myth of progress through rational mastery. The modern Left had dreamed of and in many respects gained power. In the process, the modern Left had chosen to sue for deformed, alienated, and ultimately utilitarian concerns with careerism, as full-time waged employment, consumerism, especially in relation to suburban home ownership, and fashion, in commercial popular culture. Instead, that is, of maintaining critique of the system. The modern Left had sustained an unnaturally narrow and limited instrumental conceptualization of the generalizable interest. On this understanding of his work, it follows that Habermas might well have cast about in the 1970s for some alternative that would truly represent the generalizable interest of all citizens in maintaining and expanding lifeworld conditions against colonization by the system.

Interesting for the argument of this book is that Habermas at the time found this representative in the green-left and communitarian new social movements. For him, these were indeed advocating countersystemic demands for lifeworld freedoms. The new movements had displaced the modern Left as the agent of radical Enlightenment because they more genuinely expressed the generalizable interest in freedom as autonomy. The new movements defended a truly rational and so radically enlightened lifeworld against encroachment by an instrumentally rational system dependent on truncated Enlightenment ideals. Emerging from within what remnants of the lifeworld remained in the 1960s and 1970s, the new social movements actualize the consensual ethical norms that express the generalizable interest in freedom. Habermas felt that the new social movements were serving as counterfactual political incursions that could and should eventually serve to constrain system imperatives. Although he does place the destruction of the environment or nature as amongst the most egregious pathologies of modernity, Habermas was however unsympathetic to what he saw as the new social movements' self-understanding. For him, the movements themselves may have given expression to lifeworld norms, but as had been Horkheimer, Adorno, and Marcuse, the new social movements and Greens in particular were unable to articulate a clear rationale for resisting the system. For Habermas the new social movements all too readily devolved into the early critical theorists' Nietzscheanism, the contradictory anti-humanist yet subjective "neoconservative defense of postmodernity" epitomized by Foucault, while provoking an equally confused antimodern conservatism.[20]

Insofar as its advocates sought to redesign society along harmonious lines, to decenter humanity and prioritize nature in its place, green-left modernism would remain politically impotent. The new scholars and activists misunderstood the relationship between society and nature and discounted or misunderstood the liberatory potential of the noncoercive social relations that for him characterize "lifeworld interactions." Habermas was in this way rejecting the recourse to philosophies of nature used by those he derided as green Romantics in the same terms as he rejected Horkheimer's, Adorno's, and Marcuse's critical theory. Whether left implicit by Horkheimer, Adorno, and Marcuse or made explicit by green-left modernists, each in Habermas's opinion uses the holistic idea of nature as a decentering justification for a moral voluntarist ethic that expresses benign natural processes.[21] The problem is that Habermas's lifeworld conceptualization of the generalizable interest is, albeit implicitly, also a pure inalienable *natural* state with which the new social movements simply happened, almost as an accident of history, to be in tune. In other words, Habermas's philosophy and the ideology he hoped to propagate seek to remake authority in the image of the worthy ethical agent, whose lifeways and moral choices express some inalienable truth that authority should represent and embody. As Vogel shows, this move merely pushes back a step Horkheimer's, Adorno's, and Marcuse's, as well as Nietzschean, justifications for practical action. In the guise of the human interest in communicative rationality, Habermas "appeal[s] to a noumenal (and thus *a*historical) nature independent of human action."[22] The communicative rationality native to the lifeworld prepolitically defines normative matters of practical advocacy in terms of a future-present realm of benign human interactions. Habermas yearns for a world in which ethics come first, and in which authority can be legitimately made to express ethical mores. Although this form of structural moralizing is complicated by his efforts to get beyond what he saw as the subjective moral outrage of the new social movements in favor of a dispassionate intersubjective "communicative ethics," it remains the case that communicative rationality embodies an inalienable, essential truth. Indeed, over time Habermas would abandon protest culture as the vehicle for progressive change altogether and elevate responsibility for operationalizing the counterfactual of communicative reason to constitutional and juridical deliberation, albeit urged by a robust "public sphere" in which the forceless force of the better argument would emerge organically from rational debate. With this move, Habermas would alter the future focus of critical theory away from the critique of political domination and towards the critique of flawed ethics.[23]

This criticism aside, Habermas made clear that what he saw as the Nietzschean aporia could lead only to individualistic withdrawal into sub-political communitarianism or the triumph of the ironic, cynical calculating egoist. Habermas interprets the reproduction of modernity in terms of a standoff between the advocates of two forms of practical reasoning, the inalienable truth that communicative rationality expresses and the calculative egoistic utilitarianism promoted by the system. His observation is that the modern Left's compromise with big government and big business had been mobilized by its advocates' shared acceptance of the individualistic utilitarian social contract. This exposes the modern Left's conflation of the normative demand for freedom with an analytic that appears mobilized by what, in chapter 1, I discuss as the moralized politics of fear. Upon invitation to compromise, the modern Left disavowed itself of the politicized politics of fear, which arguably had motivated mass organization in the first place, and had come to envision the expansion of freedom as simply a logical extension of profit-driven techno-scientific development.

Yet Habermas too seems to regard with suspicion an analytic based on the political view that the achievement of freedom in liberal democracy is merely a contingent product of collective détente that is ultimately oriented to maintaining civil peace and defense against enemies abroad. Arguably, this may be due to Habermas's historical observation that the tripartite compromise that had bound democracy, capitalism, and the modern Left had also effectively steered the realization of public welfare onto calculative utilitarian grounds. The pathological products of the compromise were after all a rampant and deadly, blindly ends-focused military-industrial form of capitalism, a careerist political and bureaucratic administration, and an alienated, fear-ridden, conformist, and mannerist consumer culture. With the not inconsiderable benefit of hindsight, however, Habermas seems to have taken a historical contingency for a necessary condition. Whereas the egoistic utilitarian calculus had since the early nineteenth century shaped individuals in the image of the modern bourgeois citizen, it also produced a modernist citizen, unwilling to accept the utilitarianism that the modern myth of progress through rational mastery seemed to engender. The affluence, healthcare innovations and other techno-scientific developments, access to education, work, free time, and, notably, independence from parental discipline and familial obligations afforded by the modern Left's compromise had facilitated the embrace by a whole generation of a modernist political culture that resisted the systemic vision of life as the mere egoistic calculation of costs and benefits.

The primacy of utilitarian egoistic individualism and its critical alternative, the expressive modernist individualism of the new social movements, represent two sides of the same coin. In both cases, some moralizing, essentialist truth about human nature takes priority over politicized agnosticism aimed at coercing rationalized authority into accommodating demands for freedom while primarily responding to what Williams calls the first political question. The relevant analytic distinction may not be between an undesirable system and normatively desirable lifeworld but between two benign and malignant modes of asserting moral truth, and agnostic concession to the artificiality of political authority that focuses normatively on simply contesting the decisions of holding power. Commitment both to benign and to malignant truths serves to moralize what is ultimately a political analysis. Shifting analysis onto a footing that prioritizes some, often unsatisfactory and always incomplete, degree of public welfare or civil peace emphasizes a focus on efforts to expand political freedom and, by implication, reaction to such efforts.

At this point, the dynamic that Adorno had elaborated in philosophical terms might shape the normative dimension of analysis, but as he recognized, only at a cost. The dynamic between nonidentity or the negative *critique* of domination, identity, or the *assimilation* of critique to "the system," and false nonidentity or critique's *recuperation* as both a system input and a foil to subsequent critique, implies that normativity cannot be simultaneously both critical and positive.[24] When collectives cede to a stalemate that decides the distribution of welfare in ways that expand political freedom, those collectives and the individuals who constitute them become avatars of system imperatives. Analytically, focus is directed to the achievement of détente between parties that provides for an expansion of political freedom. Adornian critical theory remains committed to exposing the negative moment at which nonidentity critiques both the identitarian and false nonidentitarian dimensions of the system. In liberal-democratic settings at least, such agnosticism is historically bound to regard the rationalized authority of the artificial state as the least worst way in which those holding power might respond to the first political question. In this perspective, ontological appeals to some benign or malignant truth about nature and human nature may at best serve as impotent moralizing or, at worst, as grounds for the reassertion of raw power against rationalized authority.

The Crisis of Democracy

As I suggest at the opening of this chapter, the crises of the mid-1970s were not only important to radicals and reformists but also to conservatives and reactionary scholars and activists. Recent historical work suggests that intellectual and practical efforts of people on the right to come to grips with what appeared from their perspective to be a crisis of democracy[25] would help to shape the forthcoming New Right. This New Right was initially funded by and gave voice to the concerns of small to medium size "family businesses" that had been nurtured since the Progressive Era of the early twentieth century. Such concerns had been amplified by the widespread popularity of the New Deal in the 1930s and in response to the Great Society programs and civil rights achievements of the 1960s. Over this period, these family business owners and managers and their representatives in commercial guilds regarded themselves as sidelined by large-scale corporate shareholder-owners and high-level managers. The latter representatives of "multinational" capital were perceived to have reconciled themselves to progressive reform policies on grounds that well-regulated industries stabilize power relations between business and organized labor. For the family business lobby, such collusion between big government and big business provided further evidence that state managed intervention into markets, whether through regulation or taxation, was launching the United States onto a slippery slope toward a corporatist or even communistic totalitarianism.[26]

By the 1960s, leaders of the family business community were arguing that, "virtually the entire American business community [has] experienced a series of political setbacks without parallel in the post-war period."[27] In the late 1960s and early 1970s, such views had started to win over large-scale corporate shareholders and some beneficiaries of inherited wealth, who felt themselves unjustifiably footing the bill for progressive reforms or harmed by the new raft of environmental regulations. Well-funded and organized groups representing interests in free markets, in concert with academic advocates of rational choice, public choice, and social choice in the social sciences and the law and economics theory in legal studies, most working in sympathetic universities associated with the Virginia school, began not only to promote deregulation and small government but also to use the media to delegitimize labor unions, civil rights groups, and the new social movements. As Frances Fox Piven, Nancy Maclean, Jane Mayer, and others observe, a wave of increased donations to right-leaning think tanks including the John Birch Society, the Heritage Foundation, the American Enterprise Institute, the Cato Institute, and the Hoover Foundation helped to fund a well-coordinated "rollback agenda, making specific proposals for shifting the brunt of taxation from capital to wages."[28]

Although initially meeting with little success, this putsch would grow and expand into a complex web of think tanks, privately funded research centers and institutes within public universities, lobbying groups, and media organizations. Over time, a New Right movement would develop a coherent philosophical and ideological program designed ultimately to constrain those regulatory institutions dedicated to deontological oversight of unfettered capitalist exchange. In their place would be promoted regulatory institutions dedicated to upholding market forces as the only justifiable and legitimate source of authority in society. The New Right in this sense sought to alter both political authority and the contours of political freedom, as constituted and expanded since the Progressive Era. It sought in fact to replace the institutions associated with artificial political authority that, in the 1930s and again since the 1960s, had proven all too amenable to radical and reformist demands for the restraint of markets and for the redistribution of profits and rents. Against such artificial authority, New Right scholars and activists would "activate an effective counterintelligentsia." This counterintelligentsia would in turn, by influencing "politicians, big business, the media and the courts," help to reestablish the natural moral authority immanent in a system of free exchanges amongst autonomous individual property owners.[29]

Influential figures such as the corporate lawyer and future Supreme Court justice Lewis Powell famously listed the setbacks confronted by American business and offered a retaliatory strategy in his widely publicized yet "confidential" 1971 memorandum to the United States Chamber of Commerce. Powell argued that both the Democratic and Republican Parties' enthusiasm for political intervention into the economy was sustaining "student radicals, antiwar demonstrators, [and] black power militants," who with the support of "the liberal elite . . . [have] turned against the depravity of corporate America."[30] Around the same time, sociologists and political scientists sponsored by the Trilateral Commission think-tank developed a similar argument. Although self-presenting as the voice of a nonpartisan group of liberals and concerned industrialists, their report itself expressed an inherently conservative response to the crisis. In the commission's view, a combination of "tax-spend-borrow" policy-making and overgenerous welfare programs, including higher education disbursals such as the Pell grant scheme, were fostering demands for an "excess of democracy."[31] Rather than envisioning social disruption as evidence that the established boundaries of political freedom were illegitimate and required shifting, the report evoked a Cold War version of the moralized politics of fear: Western liberal democracy was being weakened from within and so would be rendered unable to respond adequately to the first political question in the face of the Soviet Bloc.

The Powell memo and the commission's *Crisis of Democracy* report explicitly referred to the ways that progressive reforms had undermined citizenly faith in free markets. Both texts portray the threat as symptomatic of a taxpayer-sponsored shift toward achievement of the longstanding radical and reformist Enlightenment objectives of universalizing conditions for freedom. These had led to the normalization of forms of political and cultural modernism that were once the preserve of a rightfully marginalized counterculture. This effort to revive the moralized politics of fear—direct fear of the burden on hardworking Americans and indirect fear of moral permissiveness—was designed to alert sympathetic citizens to the dangers posed by the spread of nonconformism. If left unaddressed, the creep of leftism would encourage labor unions seeking shorter working hours in response to automation to build alliances with green-left groups seeking quality of life over material comforts. And it would encourage working-class white, black, and immigrant citizens to identify with each other as unable to enjoy fully the benefits of modernity. Such state sponsored and taxpayer funded modernism would, furthermore, popularize atheism and communistic values. The New Right traced a narrative of declension through which Western civilization had fallen victim to the unruly children of Enlightenment. Current elites could summon the will neither to discipline nor to punish radical Greens, recalcitrant peaceniks, delinquent youth, unruly feminists, and angry blacks, indigenes, and other minorities. The aim was therefore to delegitimate both the modern Left's demands for further New Deal and Great Society–type reforms, and the nascent modernist Green Left's demands for a lifeworld conducted in harmony with natural equilibriums and at the more human scale of local eco-communities. The crisis of democracy narrative was therefore simultaneously a reaction against Democratic liberal elites, patrician Republicans, and conservative corporate leaders who had sided with and in some cases sponsored "irresponsible value-intellectuals" who advocated leftist, green, and nonconformist values on university campuses and in high schools.

Initially, this movement adopted the moralized politics of fear to mobilize those whom President Nixon had labeled the "silent majority" against green-left modernist decadence and the reformist Left of the Democratic Party. Powell and the Trilateral Commission had both warned that inattention to the unfolding crisis would undermine "the strength and prosperity of America and the freedom of our people." Indeed this narrative did exert a destabilizing effect on the Democratic Party and further weakened the modern Left. By the late 1960s, the Democratic Party had come under pressure to end prosecution of the war in Southeast Asia, advance civil rights, extend welfare entitlements and access to education and healthcare, and ramp up environmental regulations while curtailing consumerism. Driven by suspicions that the above developments were problematic for their constituencies, a combination of labor unionists, civil rights campaigners, and centrist Democratic Party activists established the Coalition for a Democratic Majority to resist the new cabal of party "longhairs": former New Left and student leaders, Greens, and feminists. Insofar as its members sought to advance reform leftism by reviving modern political culture, the committee would be drawn into the orbit of the Republican Party and eventually drift to the New Right.

Meanwhile, the other issue splitting the Democratic Party was of course civil Rights. Since the 1960s, working class mainly southern whites and northern "ethnic" whites had been drawn towards the Republican Party, a process inaugurated by the populist and nationalistic yet failed campaign mounted by Barry Goldwater in the early 1960s. The Republican Richard Nixon's electoral success in the late 1960s was in this view the result of two nuanced deployments of the moralized crisis narrative. Nixon had appealed to southern working-class whites by calling for Congress to respect states' rights. Meanwhile, Nixon had sought the support of northern working class ethnic whites—the historically integrated and largely Catholic, Eastern Orthodox, and Jewish urban migrant communities—by demanding that city governments do their jobs and enforce law and order. This strategy of exacerbating class and identitarian tensions was very effective. Nixon promised a government that would curtail handouts to historically disadvantaged black and immigrant individuals as well as nonremunerative community development programs that prioritized minority interests, including countercultural anti-family issues such as women's shelters, drug and prisoner rehabilitation, abortion clinics, and gay lifestyles, and instead favor "real Americans."[32]

All the while, Nixon trumpeted his credentials as a responsible moderate by promoting mainstream environmentalism and expanding consumer protection laws. The Nixon administration famously established a "series of laws and regulatory initiatives establishing a massive pollution control apparatus at the federal level," the Environmental Protection Agency. Accordingly, the new federal bodies and pliable regulatory regime responded to that dimension of the crisis narrative which recognized that "the environmental movement itself was not in itself a viable threat to the legitimacy of the state, but the vast outpouring of discontent manifest in the numerous movements of the time was widely perceived as a threat to the state and to democracy."[33] The Right was able to drive a wedge between, on the one hand, a white working class of southerners and ethnic northerners, whose upward mobility was stalling and whose mannerist conformity supported sensible consumerist and pollution control responses to the problem of the environment, and, on the other hand, working class minorities, "value intellectuals," green-left radicals, and liberal politicians seeking to build solidarity and promote a political vision of solidarity that extended across racial, ethnic, and regional divisions. Sometimes, but not always, subtle in its appeals, this wedge sought to expose the entire Left's universalizing aspirations as costly to the constituencies that it was alleged had worked hardest to build America, and who stood to lose not only material privileges (northern working-class ethnic whites) but identitarian prestige and status (southern working-class whites).

It is against the New Right's crisis narrative that Nixon's ostensible moderateness and environmentalism in response to what, as seen in chapter 2, the president had elaborated in his State of the Union Address of 1970 as "the great question of the '70s," makes most sense. The message was that it was not going to be lazy unionists, unruly minorities, women, radical Greens, gays, peaceniks, intellectuals, or counterculturalists who would provide the know-how to make reparations to the environment, or who deserved to benefit from the hard work that making peace with the environment would involve. Following a pattern resembling right-wing reactions to social turmoil since the French Revolution, the Nixon administration invited one fraction of society to envision itself as the rightful but wronged beneficiaries of a fragile construction: a settled order endangered by radical and reformist demands for an expansion of political freedom to undeserving others. The administration's strategy was to depoliticize politics by separating the shared moral duty to "make peace with nature and begin to make reparations for the damage we have done to our air, to our land, and to our water" from the threat posed by destructive left utopianism. In this way, Nixon "preempted the environmental issue by putting forth a consumerist, technology-centered, pollution-control approach to such concerns as water pollution, garbage disposal, sewage discharges and air pollution."[34] This move isolated widespread concerns with the environment from radical green-left modernist demands, repackaging the former as the preserve of a managerial state and newly emergent service and technology sectors of the fast postindustrializing economy.

Notwithstanding Nixon's resignation in 1975, a cascade of environmental legislation followed under the Republican Ford administration. By the time of Democratic president Carter's inauguration in 1977, "public support for environmental protection had become so broadly and deeply founded that it was, in effect, now part of the national consensus—that array of issues publicly accepted as an essential and priority concern of government." The environment had been established as a political object, a constellation, subject to policy-making and contestation amongst interest groups. This historical development not only allowed the right to separate moderate responsible citizens from dangerous utopians but also mainstream from radical green groups. As Gottlieb shows, professionalization led to the "creation of a new breed of environmental organization with expert staff, especially lawyers and scientists, and a more sophisticated lobbying or political presence in Washington." The new mainstream Green movement positioned itself as the voice of nature amongst the many other interest lobbies inside the beltway. For many such Greens, it made sense to seek and accept the embrace of the state in the United States in the 1970s. Yet while it made perfect sense for moderates to work to influence a state that would in the absence of political pressure advantage business over other interests, this move cast the radical Green Left into an unusual position. Insofar as the green-left synthesis had developed into albeit an often-fractured voice for nature and social justice from without and often against the state and business interests, professionalization drove deep and social ecologists and their eco-communalist and eco-anarchist followers into a corner. Having finessed a structural moralizing position that saw them defending the same goal as the professionals—the holistic ecological harmonizing of society's participation in the ecosphere through ethical eco-aware choices—radicals could only embrace ever more extreme forms of direct action or ever deeper subpolitical communitarian withdrawal to the local scale of prefigurative eco-communities and urban squats, direct action camps, sit-ins, and blockades.[35]

Of course, the influence exerted upon the state-market nexus by mainstream Greens did not last, nor did positive political consensus on the environment. As the two waves of the so-called energy crisis increased fossil fuel prices, President Carter, having "portrayed himself firmly as an environmentalist [and] actively courted the environmentalist vote . . . ," saw his Administration almost immediately plunged "into the inherently contentious and unpopular business of regulating national energy consumption, thus setting the stage for the bitterly adversarial environmental politics of the 1980s." Informed by many of the well-funded New Right think tanks that had developed specializations in the advocacy of supply-side economics over Fordist-Keynesian macroeconomic intervention and self-responsibility over welfarist redistribution, Carter established an economic cost-benefit analysis driven Regulatory Analysis Review Group, "whose mission included reducing the cost of federal government environmental regulations." Importantly, the group's target was not only environmental regulation but also all forms of alleged government overspending. Such oversight had been initiated under the Nixon administration and reinforced under the Ford administration in response to increasing pressure from the business lobby to rein in government debt. However, it was Carter who combined his avowed environmentalism with his equally avowed small town common sense and status as a born again Christian to actively seek the support of middle-class citizens and market interests in such a project.[36]

In this regard, Carter combined elements of the New Right's response to the crisis narrative with an all-encompassing national policy platform that, pretending to rational objectivity and common-sense knowledge, would safeguard not only the environment but the national interest by simplifying all matters that came before government to the rigors of utilitarian cost-benefit analysis. By depoliticizing debate over the distribution of access to social goods in this way, Carter may well have been the first left neoliberal president. The end of the 1970s thus brought with it a large-scale shift in the balance of organized interests capable of driving institutional responses to the first political question in the United States. This shift empowered free-market fundamentalists intent on reducing the tax burden borne by business, even though dispassionate analysis suggested at the time that business actually benefits from what they call a mixed economy. Indeed, while it is "possible to see the story of the 20th century in terms of the triumph and expansion of liberalism, from the New Deal through the Civil Rights movement, feminism, the gay rights movement and environmentalism," it is perhaps more accurate to envision the story in terms of those "significant parts of the American population [that had] always dissented from liberalism" and who the well-funded New Right would successfully mobilize.[37]

The Reagan Revolution

For the New Right, things really began to come together with the election of the Republican Party nominee Ronald Reagan to the presidency in 1980.[38] The New Right had gained electoral momentum since the evisceration of the patrician reformist faction of the party following the failed Goldwater putsch in 1964, and Nixon's efforts to court the southern white working class and northern ethnic Democrats thereafter. Most noticeable about the so-called Reagan revolution was that the party had come to espouse the view that rationalized authority, represented as "government," was always and everywhere the source of, rather than the solution to, the nation's problems. The Reagan campaign trumpeted the abandonment of state planning, bureaucratic dirigisme, and redistributive social policy in favor of market stimuli, deregulation, and individual self-responsibility. The campaign argued that "getting government off our backs" would automatically lead an imperiled nation out of an outmoded and uninspiring past and into a "sunrise" world of individual entrepreneurialism, networks of independent contractors, nimble just-in-time post-Fordist production regimes, and creative, stimulating consumerist lifestyles. The silent or, now, moral majority of suburbanites and rural people who had, it was alleged, continued to work hard, maintain their patriotism, and conform throughout the 1960s and 1970s could at last be freed from the false god of collectivism and protected from the extremes of eco-terrorism. Following the path broken earlier by Nixon and Carter, Reagan and congressional Republicans deftly cast national political and economic issues—weakness in foreign policy, stagflation, energy costs, cultural fragmentation, and, importantly, marriage breakdown, petty crime, and drug use—as a morality play that only a new culture of respect for stern leadership, family values, and Christian faith could resolve.[39]

Extending the crisis narrative into the 1980s, advocates for the Reagan revolution portrayed the widespread affluence that had accompanied the three decades of high economic growth and the concomitant expansion of access to welfare and low-cost education as fostering a far-reaching moral collapse. This moral collapse had rendered many Americans unable or unwilling to confront the nation's problems. Affluence had led to a decline in the willingness to work hard and save, instead encouraging profligacy. Access to welfare had promoted laziness and inhibited individual ambition. Access to higher education had led to an intellectualization of social life, a tendency to overzealousness in the seeking out of root causes for social problems. The Reaganite alternative was to attribute social problems to human weakness and accept as natural, the Darwinian consequences.[40] Informed by a well-funded, increasingly national and coordinated network of think tanks and media outlets, the Reagan-era Republican Party confected a view of past American governments, both patrician Republican and liberal Democratic, as having incurred unsustainable levels of debt and stifling levels of bureaucracy and regulation in support of morally weak individuals who actively refused honest work. Reagan, his advocates and followers in the states and in Congress protested politicians who had indulged in the moral turpitude of the 1960s and 1970s. Such laxity had fuelled the willingness of both parties to simply print new money in response to social problems while pandering to un-American mores and interests both within the nation and on the international stage. All the while, these out-of-touch politicians had supported the interests of an effete elite concerned more with political correctness and wild nature than moral fiber and a strong work ethic.[41]

Soon after Reagan's 1979 election victory, the congressional Republican Study Committee issued a report, *The Specter of Environmentalism*. This report cautioned that the political cultural shifts of the 1960s and 1970s had indeed fostered a hysterical and overintellectualized, socialistic response to the problem of the environment. Such an overreaction was not merely stifling economic growth but, as Reagan had repeatedly emphasized on the campaign trail, the will of "the 'conservative, God-fearing free enterprise people' that had made America strong."[42] As an alternative, Reaganite members of Congress and the wider New Right networks advocated a new and sensible "grassroots" form of environmentalism that from its redoubt in the southwest and west of the country was taking on the overzealous professional Green lobby. Ironically, these grassroots political organizations, notably the Wise Use movement, depended on "the funds and support of industry, large landowners, and other powerful interests" that were often the direct beneficiaries of large federal subsidies.[43] For environmental historian James M. Turner, the Wise Use movement was not a grassroots movement at all. Rather, it was an early iteration of what would emerge as a series of well-funded "astroturf" movements. As a pseudo-grassroots movement

> wise use tracked broader shifts in the political strategies of the New Right. Elsewhere in the country, conservatives had already begun to shift attention away from old issues, such as race and states' rights, and toward positive rights-based claims of individuals [and local communities]. That is, instead of positioning themselves against something, they portrayed their conservatism as for something—and something resonant for many Americans.
>
> As outlined in *The Wise Use Agenda*, the movement aimed to reduce the scope of federal environmental regulations, expand opportunities for economic development, and safeguard individuals' constitutional rights while promoting a "New Environmentalism" that respected people and nature.
>
> This strategy, emphasizing rights, pursued through grassroots activism, and staking a rhetorical claim to the middle ground, helped align wise use with a rights-based politics nationally dominant in the 1980s and 1990s and central to the rise of modern conservatism.[44]

As such, it seems mistaken to dismiss movements such as Wise Use and the earlier Sagebrush Rebellion as simply anti-environmental, as did some mainstream and radical Greens at the time.[45] To ignore this dimension of the New Right agenda during the Reagan revolution as blunt anti-environmentalism is to miss an important shift in political culture that had taken place in the 1970s alongside the consolidation of the environment as a discursive object.

The mainstream green lobby had embraced the state-market nexus precisely at a point when people on both the left and the right had become increasingly suspicious of the Enlightenment myth of progress through rational mastery. Something had changed in American political culture by the 1980s, and

> as public lands protection grew in scope and became dominated by technical analyses in the wake of NEPA and as environmental groups faced sharp criticism during the Reagan Administration, [professional] environmentalists assumed that such empirical arguments and legal strategies, rather than personal testimony and moral convictions, were the most effective way to advance their cause.[46]

Professional Greens had wrested the idea of facilitating society's harmonious participation within the ecosphere away from the ragtag congeries of utopians on the left, from visionaries such as Foreman and Bookchin. However, simultaneously, the well-funded and relatively organized utopians of the Right, self-presenting as responsible moderates, worked to undermine the efforts of both radical Green-left modernists and mainstream reformist Greens.[47]

What the Wise Use groups and the New Right in general offered to bring to the United States was indeed a new "sunrise." This time around, however, the shades of dawn would be colored not by Promethean courage to measure, predict, order, and control but by a new kind of courage. This new sunrise would be led by the brave and worthy few, bearing the courage to exercise moral restraint in the face of the nation's boundless natural and human resources, and to use them wisely. The Reaganites and New Right would call upon citizens to embrace humility. The strategy was to combat and ridicule as hubris the ambitions of both the modern Left and Right to exert rational mastery through the political authority of the state. This was a strategy perhaps most tightly encapsulated in Reagan's response to Carter's efforts to explain universal healthcare during the 1980 election campaign: "There you go again."[48] The Enlightenment myth of progress through rational mastery had sustained an overzealous "nanny state" wherein two "un-American" qualities had combined: overzealous command and control regulation that knew no limits and immoral licentiousness. The New Right sought to revive a moribund economy and political culture on the basis of arguments that these had become too dependent on an illegitimate and ultimately unnatural political authority.

Justifying a Return to Moral Authority: The Hayekian Cosmology

Several scholars and activists may be called upon to exemplify New Right thought and activism. While economist Milton Friedman and fiction author Ayn Rand held significant public profile at the time, and economists and political scientists of the Virginia school that advanced the public choice doctrine in the social sciences and the Chicago school that advanced the ideas of law and economics worked assiduously to refashion research agendas within their disciplines, one figure in particular stands out for the clarity of his thought, philosophical completeness of his vision and public reach of his views. This is Friedrich Hayek, the so-called guru of neoliberalism and progenitor of the "neoliberal thought collective." Indeed, whereas Friedman had worked within the orbit of Keynesianism in the 1930 and 1940s, helping to establish the federal tax regime, Hayek had opposed interventionism throughout the New Deal Fordist-Keynesian era. It was as a widely read author of fiction rather than as a political actor that Rand was most influential, and her contrarian libertarianism was at odds with early New Right appeals to authority. And, while James Buchanan, Gordon Tullock, Gary Becker, Richard Posner wielded great influence within the academy, the public impact of their efforts were felt only as politicians, policymakers, and judges put the ideas of the social good as untrammeled property rights and wealth maximization into action. Indeed, it was public intellectuals such as Irving Kristol, George Stigler, Henry Simons, George Gilder, and Jude Wanniski who perhaps did most to translate Hayekian ideas into ideological arguments. The capacity of these New Right figures to influence politics, policy, and jurisprudence was arguably a consequence of the philosophical heavy lifting finessed by Hayek in the 1940s and 1950s. As with the other New Right luminaries, Hayek had, with considerable assistance from the business lobby, developed a powerful justification for rethinking the role played by political authority in organizing social relations.[49]

However, Hayek's thought is also an important place to begin insofar as he explicitly lays out the Right's challenge to rationalized authority in terms of the discourse of Enlightenment. Hayek saw himself not only as a thinker seeking to discredit but also as an activist seeking to undermine the dangerous enthusiasm for Enlightenment reason that in his view had taken root in universities, government, and the wider political culture.

Importantly, "for conservative intellectuals such as Hayek, the belief in the market went along with a kind of antirationalism that had close cousins in conservative religious circles." The ultimate justification for Hayek's worldview was that authority could not be legitimately wielded to uphold any ideal of the social good directly, but only to ensure that society would work in harmony with the "spontaneous order" of the cosmos. The arena of social relations in which this spontaneity arises is not that of organizations, such as governmental bureaucracy or legislature, but the market. This is because the spontaneous order of markets "rests not on common purposes but on reciprocity, that is on the reconciliation of different purposes for the mutual benefit of the participants." For Hayek, the spontaneous ordering that arises in and through markets had, despite setbacks due to misguided elites and crazed revolutionary leftists, produced the greatness that was the free world of the twentieth century. For Hayek, it is the hubris of bureaucratic planners and meddling legislators that risks destroying such greatness, and setting a free people onto "the road to serfdom." Essentially, Hayek argues that individual primitive freedom is both the condition for and outcome of this historical unfolding. His argument is naturalistic and depends on a Darwinian understanding of the ways that individuals, left unhindered, will spontaneously align society with the cosmological order. His worldview is committed both to ontological holism and moral voluntarism.[50]

Indeed, "what Hayek describes is entirely holistic, individuals are not considered qua individuals, but as vehicles of specific socially inculcated beliefs, as nodes in networks of social relations."[51] In the 1940s and 1950s, Hayek used the commitment to holism to justify a distinct model of authority:

> That a particular order of events or objects is something different from all the individual events taken separately is the significant fact behind the [phrase] . . . "the whole being greater than the mere sum of its parts." . . . It is only when we understand how the elements are related to each other that the talk about the whole being more than the parts becomes more than an empty phrase.[52]

Decades later, Hayek continued to maintain the same ontological commitment, arguing that

> the overall order of actions in a group is . . . more than the totality of regularities observable in the actions of the individuals and cannot be wholly reduced to them. . . . A whole is more than the mere sum of its parts but presupposes also that these elements are related to each other in a particular manner.[53]

For Hayek, it is plain hubris to pretend that policymakers or, worse, politicians could design policy as if the entire social order were accessible to human intelligence, and then rationally plan to redistribute access to social goods along lines that run counter to those naturalistically presupposed by free market exchange.

Hayek's argument is of course a casuistic justification for the authority of those who in fact do know that the whole is greater than the sum of its parts and always tends toward a satisfactory order. Authority must be wielded on the basis of an intuition held by those who benefit most from a cosmological ordering force that can be cruel but is ultimately beneficent in the long run. This ontological commitment sustained Hayek's "holistic rhetorical strategy for *laissez-faire*." Such a rhetorical strategy held out great appeal to "American business conservatives [who increasingly over the 1960s and 1970s had] viewed the development of free market philosophy with great interest, recognizing it as an elegant, sophisticated statement of their worldview." In particular, business was attracted to Hayek's "critique of the wholly artificialist conception of legislation," and rejection of any justification for "the omnipotence of the legislative power." For Hayek, the legislative arm of government was fundamentally designed to disrupt markets' nature-like capacity to produce and distribute social goods in ways that would otherwise be attuned to the cosmological order. The artificialist conception of legislation is problematic for Hayek because it legitimates misguided, often resentment-fueled, demands for some alternative nonmarket source of authority over the production and distribution of social goods. He regarded legislative interference as artificial because an authority that pretends to deontological status in relation to free markets is self-evidently unjustifiable. For Hayek, the tasks of government should be limited to promoting what he labeled the rule of law but which is more accurately defined as inherited common law, not legislation or what would be in Roman law societies the civil code. The Hayekian rule of law expresses a historically arrived at concept of justice that has "been gradually developed, just like languages and money, on the basis of repeated experience of their infringement." Insofar as in common law, the "selection of rules of just conduct is the source of [further] social progress," this is an argument for a return to premodern conceptions of law, albeit one aware of the Darwinian pressures of natural selection. Indeed, Hayek's argument for the primacy of the rule of (common) law echoes Carl Schmitt's and, ironically, V. I. Lenin's totalitarian understandings of law as that form of authority in which wise judges merely channel and apply the community's accumulated moral wisdom. For Hayek, in contrast, legislation is "'enacted' or 'constructed,' and in this sense constitutes a 'manufactured' or 'artificial' order" that, grounded in hubris, is doomed to undermine social progress.[54] Hayek thus offered to a willing audience of well-funded New Right actors a justification for working to depoliticize rationalized authority's capacity to respond to the first political question.

At the height of the Reagan revolution, it was the thought leader Irving Kristol who summarized Hayek's aims to restrict political interference in markets by promoting supply-side and denigrating regulatory demand-side policies. Policy focused exclusively on the supply side was represented as

> originat[ing] in deliberate contrast to the prevailing Keynesian approach, which emphasizes the need for government to manage and manipulate—through fiscal and monetary policies—aggregate demand so as to maintain full employment. . . . [New Right] supply-side economists say government cannot really do this, no matter how many clever economists it hires, but that if business enterprise is permitted to function with a minimum of interference, it will invest and innovate, so as to create the requisite demand for the goods it produces.[55]

This is a Hayekian argument. Kristol is demanding humility in the face of an unknowable universe that is governed by natural forces, which may be channeled but not tamed. What makes legislative-regulatory intervention by political authority so problematic in this view is not that it stymies growth but that it contradicts the evolutionary cosmological forces that would exert a spontaneous ordering effect on society. As Hayek himself had argued,

> it would . . . be wrong to conclude, strictly from such evolutionary premises, that whatever rules have evolved are always or necessarily conducive to the survival and increase of the populations following them. We need to show, with the help of economic analysis . . . how rules that emerge spontaneously tend to promote human survival. Recognizing that rules generally tend to be selected, via competition, on the basis of their human survival-value certainly does not protect those rules from critical scrutiny. This is so, if for no other reason, because there has so often been coercive interference in the process of cultural evolution.[56]

To Hayek and his followers, "coercive interference" by political actors or indeed representatives in the *otherwise* natural workings of the market is not only normatively undesirable. Coercive interference is also ontologically wrong because it is based on the dualistic fallacy that humanity is the subject of history rightfully exercising rational mastery over an objective nature. That is, based on the hubristic Enlightenment assumption that the world can be measured, ordered, predicted, and controlled by, for example, politically influenced legislation or government-hired "clever economists." Legislative intervention is a product of Enlightenment's hubristic "naïve rationalism" and can only have the unintended yet wholly predictable consequence of undermining the "survival-value" of the historically accreted, market-based inherited common rule of law.[57]

This is the "law of unintended consequences" that Kristol and his colleagues consistently argued would accompany any legislative effort to address the production and maldistribution of social and environmental harms. In the view popularized by Kristol, expounded for policymakers in conservative journals such as *Commentary* and *Public Interest* and bowdlerized by sympathetic media outlets for mass consumption, the New Right favored just such a humble perspective on policy. Such humility

> look[s] at the economy from ground level, as it were—i.e., from the point of view of the entrepreneurs and investors who are identified as the prime movers. Keynesian economists look at the economy from above—from the [dualistic] standpoint of a government that is a *deus ex machina*, and which, in its omniscience, intervenes discreetly to preserve a harmonious economic universe.[58]

In Kristol's hands, Hayek's holistic ontological commitment to the spontaneous order of the cosmos is transformed into a justification for the worldview of the "prime movers," the humble yet courageous "entrepreneurs and investors" who work at the "ground level." The New Right in this way denigrated the dualist ontology presumed by Keynesian economists and state interventionists. Whether genuinely radical and reformist "socialists" or erroneous conservative patrician elites, interventionists operated on what Hayek understood as the mistaken ambitions of "taxis." For the New Right, interventionism drew on the fallacious belief that political actors in charge of a state could taxonomize: measure, order, predict, and control the natural forces at play in what is really an all-encompassing whole "cosmos." Ambitions to control the state to manage the economy, maintain full employment, and bring income into a more egalitarian distribution, or to minimize or undo environmental harms, are not only misguided and futile but run contradictory to the cosmological order. With Hayek, holistic "naturalism and voluntarism are put on an equal footing."[59] In effect, Hayek conjures an understanding of the cosmos as having meaning in itself, as an ontological feature: the spontaneous order arising in and through markets that over history gave rise to the free society. The Hayekian cosmology offered the New Right a justification for pursuit of an idealized authority that would facilitate only those laws that express the ordering force of an evolutionary process.

As such, Hayek and his followers champion a deeply individualistic, self-directing moral voluntarism. Because the state is not an artificial *deontological* entity but part of the fabric of the universe alongside everything else, those in charge of the state have been foolish to intervene in ways other than to facilitate the natural Darwinian selective mechanisms that ensure human survival and prosperity within the cosmos. Indeed, Hayek's

> holistic rhetorical strategy of *laissez-faire* says that, granted that order is emergent at the macro level, and the qualities of macro level entities are not a simple reflection of the qualities of the micro level substrate, *nevertheless* we can be confident that outcomes are desirable because there exists some mechanism, a black box, which ensures that they are.[60]

It follows that there exist wiser individuals than those interventionists currently running the nation. These are courageous entrepreneurial individuals who possess the appropriate mix of courage with humility when confronted with a universe that exceeds human capacity to know and order.

Ironically, it is precisely full and confident knowledge of this quality of the cosmological order that allows the worthy to exercise such courage and moral restraint. Hayek's arguments are, ultimately, both casuistic and "Panglossian in the social sense: [insofar as for him,] the institutional structure we inherit tends strongly to be desirable and attempts to improve it by conscious collective action are very much to be avoided." At the level of advocacy, therefore, Hayek's ideas are not so much neoliberal or even conservative but reactionary. Hayek himself explicitly justified restricting the task of political elites to what Adam Smith had described as the art of catallaxy: managing a state that upholds only the "three fundamental 'laws of nature'—private property rights, freedom of [individual] contract and the duty to compensate for damage due to fault."[61]

The aim of the state is to facilitate the spontaneous ordering principles of the universe, which will most effectively provide conditions for progress. This approach to authority as the depoliticized exercise of a management function depends upon the courage and humility of those holding power. Tendentiously, elites confront a cosmological order that cannot be understood in its entirety but which nevertheless requires unshakeable faith on the part of those elites that their own knowledge of that order's true functionality is correct. On the surface, the desired outcomes will emerge when authority intervenes in markets only to undo artificial impediments to efficient functionality and so ensure what is ultimately cosmological harmony. Beneath the surface, what is required is a return to moral authority, wielded by the righteous to ensure that society expresses Darwinian natural forces. This is of course a justification for authority on the model of enactment moralizing. Such a combination of tendentious reasoning with a Panglossian morality tale—expressing the desires of the beneficiaries of the arrangements that the theory champions—made the New Right's ideological response to the crisis of democracy extremely appealing to financial, shareholding, and rentier interests, as well as to both family business and corporate owners and managers. The crisis of democracy was not so much a crisis for democracy but a crisis for those believing themselves wronged by history.

Overcoming the Right's Paradox of Freedom

In practice, Kristol and his early New Right allies were confronted by the Right's own version of the paradox of freedom. The devil-may-care attitude toward moral laxity that capitalism seemed to require and even foster did not appeal to the broad moral majority of culturally conservative southern and ethnic white voters drawn to the Republican Party since the early 1960s and who had, it was believed, elected Reagan. Hayek had decades earlier identified the conflict between the liberal norm of individual autonomy, which sustains the entrepreneurialism essential to economic growth, and "the general observance of [moral authority, which] is a necessary condition of the orderliness of the world in which we live." Since the early 1970s, Kristol had sought to deflate such a paradox by synthesizing Hayek's arguments against state planning and for free markets with political philosopher Leo Strauss's critique of the displacement of classical virtues by instrumental individualism. For Kristol, it was capitalist faith in the Enlightenment myth of rational mastery that had, through an irony of history, fostered the emergence of a deeply irrational political "counter-culture" that was rejecting the "philosophical presuppositions of modernity."[62] In his view, "what we call the 'environmentalist' movement," is in fact emblematic of the problems that a free society confronts:

> Over-zealous environmentalists do not want to be shown [cost-benefit analyses of different approaches to getting clean air and water]. They are not really interested in clean air or water at all. What *does* interest them is modern industrial society and modern technological civilization, toward which they have profoundly hostile sentiments. When they protest against the "quality of life" in this society and this civilization, they are protesting against nothing so trivial as air or water pollution. Rather they are at bottom at bottom rejecting a liberal civilization which is given shape through the interaction of a countless sum of individual preferences.[63]

From the purview of the New Right, the bogeyman of the green-left new social movements and wider counterculture had displaced the modern Left's—noble but nevertheless to be opposed—concerns with material redistribution of the spoils of industrial civilization. In its place had arisen an ignoble mix of antinomian nihilism, communitarian refusal, and irreligious new age spirituality: modernism. Echoing Habermas's and Offe's critical theory informed analyses, Kristol argued that while the modern Left could be swayed by cost-benefit analyses, the modernist Green Left rejected rationality in favor of a greater goal: "the authority to create an 'environment' [that] will be a society where the rulers will not want to 'think economically' and the ruled will not be permitted to do so." In Kristol's view, "the identifying mark[s] of the New Left are its refusal to think economically and its contempt for bourgeois society precisely because this is a society that does think economically." As Habermas and Offe had argued, what the new green-left movements were demanding was an end to a world in which fear, pain, and physical or symbolic destruction had proliferated and for the creation of a new modernity in which integrity, recognition, and respect for socio-ecological harmony prevails.[64]

To Kristol, the green-left modernism embodied by the new social movements and the broader counterculture upon which such sensibilities drew and to which they contributed were products of capitalist success. Kristol finds that it was this success that had, ironically, precipitated the crisis of democracy. For Kristol, recognizing such an irony required that the Right demand a return to strong moral authority, preferably in the guise of organized religion, because

> if you believe that man's spiritual life is more important than his trivial and transient adventures in the market place, then you may tolerate a free market for practical reasons, within narrow limits, but you certainly will have no compunction in overriding it if you think the free market is interfering with more important things.[65]

The problem for the Right was therefore a strategic and not a philosophical or ideological one. The Panglossian narrative needed to be extended in a way that would deflate (resolve?) Hayek's paradox, and so defuse family business and moral majority concerns with the imminent collapse of the established order.

By the early 1980s, efforts by others to synthesize Hayekian reaction with Straussian conservatism would adequately deflate the paradox and provide the New Right with a strategy for electoral victory that would succeed for decades to come. Kristol's ally Daniel Bell had elaborated clearly the tension between traditional conservative desires to reestablish moral authority in order to constrain modernism and plainer reactionary desires to prop up the system that delivered to the deserving their power and privilege. In response, and with a high level of media fanfare, academic economist Arthur Laffer and Republican Party consultant Jude Wanniski made the case that simultaneously cutting taxes on the wealthy and welfare to the indolent would be the single most effective way to boost economic growth. Another New Right thinker and activist, Goerge Gilder, would urge Republican politicians to rescue capitalism from Democratic Party pro-welfare elites and radical green-left insurgents. Gilder made the moral case for supply-side economic policy that, because not impinging upon and in fact extending individuals' freedom to choose, would provide an "assurance of justice" for free markets.[66] The argument was that only a combination of welfare retrenchment and tax reductions targeting those individuals best equipped to survive and prosper, to maximize their own happiness in the Darwinian world of the market economy—the entrepreneurs and investors that Kristol had identified as economic prime movers—could revive economic growth, while disciplining those weaker individuals who had succumbed to nihilistic hedonism. Indeed,

> Gilder's project . . . was one of reshaping the capitalist moral paradigm, to produce for it a theology that justified and explained inequality whilst disrupting the middle-class flabbiness that Kristol and [ally Norman] Podhoretz saw in the counter-culture.
>
> What Gilder did in his best-selling book, [*Wealth and Poverty*] a favorite of Ronald Reagan, was to redefine the moral paradigm of capitalism; bourgeois virtues are abandoned and the willingness to take risk and embrace *fortuna* and Hayek's game of *catallaxy* are presented as the new moral benchmark.[67]

The moral failings of modernism so problematic as symptoms of capitalist society for Kristol are inverted in Gilder's argument and become sources of glory, prestige, and moral worth. For Gilder, it is the superior qualities of the wealth creators that United States society should value. Liberal thinkers once lionized by free marketers—Hobbes and Locke, Smith and Bentham—according to Gilder, could no longer provide the insights necessary for successful justification of the capitalist system in postindustrial society.

Rather, through a lens crafted by his mentor Ayn Rand, Gilder drew upon Nietzsche to argue that the interventionist regulatory welfare state "compromise" with the modern Left "merely protects and encourages a slave morality. . . . The poor, instead of being animalized by state hand-outs should be given a legalistic framework that encourages them to become risk-taking entrepreneurs." In Gilder's Nietzschean universe, employees too are morally inferior, afraid to take on the heroic risks of the entrepreneur and innovator and merely satisfied with working for another, hiding behind an interventionist state that sustains minimum wage and social security legislation. For him, "Socialism is an insurance policy bought by all the members of a national economy to shield [the weak and timid] from risk."[68] In short, Gilder rejects out of hand the possibility that, good or bad, the economic situation of any given individual might be the product of historical and cultural circumstance. Wealth and poverty, status, worldly success, and power are reconceived as reflections of an individual's moral worthiness.

No longer is it parsimony, caution, prudence, and the frugality of the Protestant ethic that are to be promoted and managed from within the utilitarian panopticon. In their stead to be monitored are qualities of risk and extravagance: willingness to live on the edge, to explore the boundaries of existence, to seek out entrepreneurial and investment opportunities. Authority should be focused upon advantaging such courageous yet—because resigned to the Darwinian struggle—humble individuals who seek to make their way in postindustrial modernity. In the New Right's emergent modernist worldview, it would be just such qualities that characterize morally worthy actors and that deserve support from the night-watchman state. This is a dual logic that, on the one hand, supports a strong moral authority that can impose rules on an unwilling citizenry softened by decades of left and liberal welfare reforms while, on the other hand, can enact and enforce only those rules that enhance the chances that the morally worthy will prosper.[69]

By channeling such a justification for authority into the moralized politics of fear, the New Right faction of the Republican Party grew increasingly effective over the course of the 1980s. New Right activism would, in Phillips-Fein's summary of recent work in political history,

> emerge from the successful, prosperous heart of the country—not from disaffected or marginal people but from those who are firmly within the American center. The movement [at the time had] not been driven by populist working-class

> resentment and anger as much as by the self-confident complacency of the well-to-do and their desire to protect their vision of the good society from the myriad threats they fear that it might face.[70]

Contemporaneous observers such as Mike Davis also saw the New Right's combination of fatalistic resignation before the Darwinian cosmological order and concomitant prioritization of moral voluntarism over collective action as appealing to a "revanchist middle strata" of affluent yet socially isolated, technically proficient yet culturally philistine, privileged yet precariously so, and predominantly white or white ethnic, suburban, exurban, and semirural citizens.[71] The New Right's moralized politics of fear represents a reactionary concession to the establishment of modernism as the prevailing political culture. Seeking to regain influence in response to the crisis of democracy, the New Right at first regretfully but increasingly enthusiastically embraced political cultural modernism.

In effect, what the New Right recognized was that the most significant and effective groups within the population were no longer those appealed to by the modern Left and the patrician Right. Rather, political actors needed to appeal to those disillusioned members of the entrepreneurial middle and aspirational working classes, small- to medium-sized family business owners, sunbelt infotech, defense, and independent trades and engineering contractors, "mom and pop" real estate investors, low- and mid-level managerial staff, and small-town and exurban professionals. The New Right's audience was made up of citizens who felt it incumbent upon them alone to survive and prosper. Of course, many such citizens held a tenuous grip on their status and wealth and lived saddled with debt and the drudgery of long hours and arduous commutes. This sociological fact perhaps contributed to the resentment stoked by Reagan and Nixon before him, fueling beliefs that "real Americans" were unjustly bearing the brunt of the taxation and regulatory burdens, all the while working hard yet standing still.

What the right had come to recognize was that the archetypal postindustrial citizen was a thoroughgoing modernist, willing or not. The Reagan revolution in this sense marked not the Right's rejection of the '60s but its embrace. Since the crises of the 1970s, modernism had continued to spread beyond university campuses, urban bohemia, and green-left circles out into suburbia and even into religion in the form of the highly popular "prosperity gospel." By propounding a quasi-biblical combination of courage with humility and grounding the concept of justice in malign natural forces, the New Right defended free markets and a primitive conception of freedom. In short, the New Right recognized that the legitimation dynamic had shifted. Authority would no longer be compelled to entertain demands from the modern Left. At the same time, radical green-left modernist demands for quality of life and sustainability could safely be ignored as utopianism, or transmogrified through supply-side economic policy into the politically neutralizing revealed preferences of "new economy" consumers for luxury, green, and ethical goods and services.[72] And even the most modest of welfare state and environmental proposals could be denigrated as the products of elite condescension. As Fox Piven presciently suggested,

> the alliance of business and religious populism [that began to take shape in reaction to the "crisis of democracy" narrative in the 1970s] is . . . bound together by a newly hegemonic doctrine that regards the operation of market forces as the workings of a divine providence and treats government measures that moderate market processes as moral trespass.[73]

New Right efforts to expose as a failure the welfare state and overzealous regulation and to characterize the poor as moral failures depended on the same combination of holistic ontological commitment with moral voluntarism that had characterized green-left modernist attempts to build momentum for radical change only a decade earlier. The New Right regrounded the holistic critique of Enlightenment dualism in a reactionary project. Yearning for panoptical omnipotence, well-funded and organized New Right scholars and activists set to recasting individuals in the image of a strong moral authority. In both cases, the naturalistic fallacy is relied upon to justify the normative goal of extending the Darwinian cosmological order. For the New Right, authority is cast in enactment terms to ensure that freely choosing individuals engage in competitive struggle. For radicals and reformists, the upshot is the loss of justifications for building and maintaining collective power sufficient to coerce artificial political authority to respond to the first political question in ways that benefit the least well off. Modernism, as a way of being in the world, provides little grounding for practical reasoning oriented to a flawed yet least-worst case of public welfare. From the 1980s onwards, it was a modernist not a modern political culture that would shape politics in the United States. Such a transformation in political culture would play to the advantage of the right for decades to come.

Chapter 4

Globalization, Neoliberalism, and Neocommunitarianism

Left neoliberalism, A Win-Win-Win Solution

Informed by polling that suggested widespread public dissatisfaction with the Reagan administration's "de-regulatory" economic policy and its anti-environmental and sometimes blatantly pro-business policy decisions, and in an atmosphere tainted by the New Right's increasing use of the news media to attack the poor and celebrate greed-is-good philistinism, Republican candidate George H.W. Bush had campaigned for the 1987 election as a moderate. Indeed, Bush had staked a Nixonian "rhetorical claim [to the] middle ground," promising a presidency that would

> seek to "improve, not roll back, environmental protections," acknowledge the importance of environmental issues, emphasize commitment to environmental protection, and promote . . . commonsense, forward-looking, and local solutions to environmental problems.

In one admiring account, while serving as Reagan's vice president, Bush had in fact been a "compassionate conservative" and "an environmental advocate 'hiding in plain sight' of Reagan's more zealous de-regulators." In office, however, while Bush self-represented in these terms, his administration favored fossil fuel interests over environmental protection, while portraying even mainstream Green demands as evidence of the broader movement's radical leftist ambitions. Soon crippled by recession and the ambiguity of his justifications for prosecuting the "First Gulf War," Bush lost office in the 1991 election to a Democrat with albeit loose ties to the 1960s New Left.[1]

Bill Clinton, with running mate Al Gore, promised that a "New" Democratic Party government would restore wage and employment levels, boost the quality of services and infrastructure, and revivify the environmental regulatory regime by embracing the newly minted international goals of sustainable development. On the hustings, Clinton and Gore set out to bridge the chasm between American values, care for the environment, and what both represented publicly as necessary market reforms. Clinton and Gore would assemble the first baby boomer administration. Led in Congress by the so-called Watergate Babies—reformists opposed to racially tinged agrarian populism and Cold War militarism—these New Democrats embraced what it called the third way between the regulatory state and free markets, an approach to electoral politicking that would later be labeled left neoliberalism. Central to this third way was a policy platform that drew on hitherto radical demands for an ecologically harmonious society that included the devolution of power from government to individuals and communities. Indeed, Clinton and Gore made much of both reinvigorating domestic environmental regulation and reviving American commitments to international environmental policy, notably ratifying the Kyoto Protocol. For mainstream Greens, by now fully established inside the Beltway and somewhat hardened by the experience of the Reagan and Bush years, the Clinton administration promised a "new beginning." While some were suspicious of Clinton, peeved by his efforts to distance the Democrats from organized labor, radical green-left scholars and activists by and large welcomed an end to Republican incumbency. It was hoped that the administration's strong commitment to albeit technocratic environmentalism would be tempered by the New Democrat's emphasis on greater citizen and community involvement in consensus-driven and deliberative approaches to policymaking and implementation.[2]

In this manner, advocates of the third way not only pioneered left neoliberalism, but also presaged what policy scholarship identified at the time as a wide-reaching, global shift from "government to governance." Led by the United States in abandoning big, centralized, top-down government in favor of lean, diffuse and collaborative governance, the post–Cold War liberal democracies would not only imbricate themselves seamlessly within the global political economy but also serve as models for a planetary policy agenda, the so-called Washington Consensus· Published to coincide with the 1992 election victory, the Democratic Leadership Council's *Mandate for Change* report outlined the Clinton-Gore "blueprint" for governing in a way that "expands opportunity, rewards responsibility and fosters community," premised on the assertion that "the best way to protect the environment is to give firms and individuals a direct and daily self-interest in doing so." The third way governance philosophy would offer a collaborative rather than conflictual approach to authority that its proponents believed offered an electoral bridge "between those who said government was the enemy and those who said government was the solution." Just as the New Right emerged from multinational business efforts to address the "crisis of democracy," the third way was part of a global response by electoral left parties to the crises of the 1970s and the chronic political economic instability that followed. As did other center-left parties across the West, the New Democrats sought to rekindle electoral support by reconceiving the modern Left's task of universalizing freedom as one of partnering with, rather than opposing, market interests. What is interesting about the electoral reformist party's third way, however, is its reliance on ideas drawn from some of the very same think tanks and economic forums that had been funded by supporters of the New Right since the 1970s.[3]

Anchoring the New Democrats' iteration of this global phenomenon were two linked beliefs about the failings of "old" Democratic Party efforts to sponsor a welfare state. First was that the state could not and therefore should not legitimately function as an institution that operated on principles different to those operating in ideally free markets. The state was to be represented publicly as a partner in developing the "wealth of nations" rather than avowed as an artificial institution standing outside and over the realm of market exchange. In short, the state was reconceived in holistic terms, as the facilitator of natural forces that may be channeled but not productively tamed. Rhetorically, this shift in attitude required that Clinton rebrand the New Deal and Great Society programs as remnants of a bygone era. In that bygone era, a misguided combination of hubris at what government could do and paternal condescension towards those in receipt of welfare had prevailed. In contrast, the New Democrats would embrace humility and respect for individuals. The third way sought to govern through consultation and consensus-driven negotiation with societal stakeholders, each equally exposed to market forces and bearing responsibility to prosper. The third way was represented as an ongoing, mutually beneficial, positive-sum policy process, rather than as the means for achieving some politically defined set of end-conditions.

Second, and it follows, was the third way belief that the legitimate goal of a government when seeking voters' approval or when controlling the state should no longer be to redistribute access to outcomes. Rather, a governing party should restrict itself to facilitating equal opportunities. Ironically, this meant embracing the combination of naturalistic holism with moral voluntarism that was simultaneously underpinning New Right efforts to combine neoclassical economic theory with reactionary political philosophy. Clinton, Gore, and other New Democrat thought leaders such as Lester C. Thurow and best-selling reformist authors such as Barry Bluestone, Bennett Harrison, and Hazel Henderson accepted as an article of faith that individuals—prior to contributing to deciding the rules that govern them—naturally seek to "truck and barter" freely amongst themselves. This natural tendency would therefore be deformed or perverted by state efforts to redistribute outcomes. State interference not only created "moral hazards" for the individuals targeted but sapped the "will of a people" to drive the economic growth from which all would automatically derive benefits.

The left neoliberalism that took shape within the Democratic Party under the banner of the third way does, however, differ from New Right ideology and what I discuss in chapter 3 as its normative justification in a recognizable way. Insofar as left neoliberalism sought to universalize access to freedom by universalizing opportunities to prosper, the New Right maintained that individual opportunities to prosper should accrue only to those who prove themselves deserving, through demonstrative acts of hard work and patriotism, for example. On such grounds, the Clinton-Gore New Democrats justified large-scale withdrawal of the state from efforts to redistribute the spoils of social reproduction as reformist efforts that would benefit individuals by virtue of their equal status as citizens. That is, the New Democrats withdrew from mass-movement politics mobilized by the politicized politics of fear. In place of such politics, the New Democrats promoted a modernist vision of personal failure to prosper as an autonomous individual. That is, the third way mobilized the moralized politics of fear in universalist terms.

Each citizen, conceived as having been granted an equal opportunity to prosper, would receive assistance in the form of "inputs" designed to facilitate their individual contributions to the growth of an "environmentally sustainable" national economy. Such supply-side policy-making would include retraining programs for green and tech jobs, and workfare programs requiring welfare recipients to demonstrate their willingness to work. Clinton's wager was that citizens would agree that ever further rounds of privatization, marketization, and free-market reregulation were the best way to facilitate a universal, therefore race and class blind, "opportunity society." The third way offered a policy frame that eschewed the modern Left's conception of citizenship as a source of political solidarity and that could assert a set of socioeconomic rights against market interests, by withholding labor, for example. The state was reconceived as one among many institutions capable of managing market forces, while citizenship was reconceived along moral voluntarist lines as a matter of self-responsibility for operationalizing a stake equally held by everyone in society.[4]

In this view, the New Democrats reconceptualized the commitment to universalizing freedom and how this commitment could and should be operationalized by a nominally left party-controlled government. In effect, this meant abandoning commitments to regulate markets from the outside, deontologically, in favor of regulation designed to facilitate the natural workings of markets from within. Insofar as the third way was premised on the support of citizens who were encouraged to see themselves as naturally autonomous actors, Clinton, Gore, and their followers evoked modernist rather than modern norms and values. Legitimation would no longer be sought as a matter of direct political contest over a party's capacity to capture the state and deliver on redistributive promises, such as for higher wages, social security, or social and environmental regulation. Rather, legitimation was sought based on direct and indirect media contestation with the Right, over which party could best facilitate opportunities to prosper within an implicitly globalizing order. In government, the left neoliberal response to the first political question abandoned reformist efforts to represent itself as the agent of organized solidary citizen movements, and in its place sought to represent atomized individuals through technocratic rational management of the economy.[5]

The third way's left neoliberalism was an avowedly technocratic response to what had been hitherto addressed as democratic problems requiring a governing party to represent partisan citizens' declared will, hence the early moniker Atari Democrats. The approach was to use state power to foster citizens' egoistic calculation of costs and benefits while promoting, in the manner of a life coach, their engagement in competitive struggle to expand, grow, and green the economy. Domestically, the Democratic Party and the president himself became managers of "workfare, not welfare," a "hand up, not a hand out," and "green growth." As the Clinton administration would make clear, the governing party was to be understood as a managerial organization, an extension of and vehicle for the nature-like forces of globalizing capitalism and liberal democracy. Extremely important to the administration's early successes was Clinton's personal ability to represent third way left neoliberalism as a race and class blind, universalist project. As Lily Geismer argues, the *Mandate for Change*

> was intended as a "guide to the progressive ideas and themes that energized Bill Clinton's winning campaign" and an outline of "a new governing agenda for a new era in American politics." Animating that agenda were several core principles: "economic growth generated in free markets as the prerequisite for opportunity for all," "equality in terms of opportunity, not results," and a rejection of both the [modern left] "liberal emphasis on redistribution in favor of pro-growth policies that generate broad prosperity" and the "Right's notion that wealthy investors drive the economy." Clinton himself . . . praised the authors' "new governing philosophy based on opportunity, responsibility, and community" and efforts to move "beyond the old Left-Right debates of the past" and depicted a Democratic coalition "pulled asunder" in the late 1960s "over issues of race, war, and cultural alienation," . . . arguing the New Democrats' philosophy and strategy had brought the Democrats in from "the wilderness."[6]

Couched in the language of modernist universalism, and contrasted directly with the New Right's thinly veiled and increasingly shrill, race inflected "dog-whistles" to southern working class and northern urban ethnic whites, the third way breathed life into a left neoliberal ideology and governance-based policy agenda that would guide the Democratic Party over ensuing decades.

By basing the ideal of a meritocratic individualism on "equality of opportunity, not results," the party could appeal to the growing strata of technocratic suburbanites, who perceived competitive markets as the best mechanism for advancing social and environmental ends, as well as to working and lower middle classes seeking reemployment from old heavy to new service industries. This said, the technocratic stratum had been the beneficiaries of an educational system presided over by progressives and shaped by the integrationist movement. Arguably many were themselves the children of working or lower-middle-class families who had experienced the American dream during the long upturn of the mid-twentieth century. Unsurprisingly, then, this group was conceived by the party as recognizing the ethical benefits of a broadly Enlightenment-oriented public education as well as the allegedly morally enervating consequences of bureaucratized welfarism for the poor, mostly black and Latino, citizens who the upwardly mobile had left behind in the era of white flight from urban cores. Coming of age during what I highlight in chapter 1 as Radkau's first great eco-boom in the early 1970s, this technocratic suburbanite "new class" was understood to be acutely sensitive to the global-local quality of environmental and quality-of-life issues. And, as Enzensberger and Luke made clear in the 1970s, were also seen to regard the bureaucracy and command-and-control regulation upholding the welfare state and environmental protection apparatus as stifling individual creativity, entrepreneurialism, and innovation and otherwise thick, local community bonds.

Indeed, prior to Clinton's election, Luke had also engaged work by New Right figures Kristol and Bell to suggest how this new class constituency might reshape the terms on which both the electoral Left would need to engage citizens into the future. He argued that an economic system premised on ostensibly clean high-tech rather than dirty heavy industries would require a technocratic class of well-educated knowledge workers and simultaneously, the mass retrenchment of working–class blue-collar and lower-middle-class clerical workers. In Luke's view, postindustrial globalization required that the electoral Left make once radical green-left quality-of-life considerations central to any program for "green growth." Luke shows that the New Democrats early on saw that "a post-industrial informationalized society [would need to be packaged electorally as] an environmentally sound economic order."[7] Electorally, left neoliberalism would need to engage deeply with hitherto politically radical and countercultural, deeply individualistic, and eco-aware modernism. Drawn into electoral politics in the postindustrial society, such radicalism would be transformed into system sustaining artificial negativity. Presciently, Luke foresaw that an electorally successful Left would need to mobilize rhetoric

> promoting a hybrid package of structural unemployment, volunteerism, soft energy paths, frugality, voluntary simplicity, decentralization and local actionism to meliorate the deteriorating situations of the "technologically obsolescent" and "technologically superfluous" classes. . . . [Left neoliberals would need to repackage a] new materially-deprived era as a "quality of life" revolution that is morally desirable. . . . "Voluntary simplicity," "frugality," "ecological lifestyles," "conspicuous conservation," "small is beautiful," "conserver society," or "simple living" [will emerge as] the new master codes of mass consumption in the age of informational capitalism, which produces less goods at higher costs for fewer people in specialized markets.[8]

Against the backdrop of large-scale deindustrial retrenchment, the "New" Democratic Party would need to channel the desires of technocratic suburbanites and urban gentrifiers for quality of life and sustainability alongside policy to ameliorate the frustrations of the newly unemployed and involuntarily independent working and lower middle classes.

This is precisely what the *Mandate for Change* sought to do. The Clinton-Gore administration would combine "liberal stances on foreign policy, civil rights, feminism, and especially the environment, with a commitment to stimulating entrepreneurship and private-sector growth," all the while promoting local community economic development as the vehicle for equality of opportunity based on a meritocratic utilitarian conception of earned individual rights and status. In such terms, Clinton and Gore maintained the cautious endorsement of mainstream Greens and, arguably, their supporters and allies amongst the technocratic suburbanites and urban gentrifiers while simultaneously enlisting all working- and lower-middle-class citizens in the cause. Clinton and Gore sought to "tackle environmental problems in a new and more economically friendly way" by proposing an "environmental-economic symbiosis—a relationship in which success in both areas is inextricably linked."[9] The newly incumbent president thus

> earned praise from environmental groups when he began speaking out forcefully against anti-environmental policy decisions of the Republican Congress, for his efforts through the President's Council on Sustainable Development to encourage new ways

> to reconcile environmental protection and economic development, and for his "lands legacy" achievements. . . . [And, for opposing New Right claims] that the jobs-versus-environment debate presented a false choice because environmental cleanup creates jobs and that the future competitiveness of the US economy depended on developing environmentally clean, energy-efficient technologies . . . incentives, and infrastructure projects to promote such green technologies.[10]

Clinton and Gore offered the public a genuine third way, an alternative to the Republicans' jobs-versus-environment rhetoric and thinly veiled evocations of "downwardly mobile" white working- and middle-class resentment.

The third way's left neoliberal vision for greening economic growth was grounded in the emergent global concept of "ecological modernization." The administration sought to replace the negative jobs-versus-environment debate with one based on the positive idea that robust economic growth could be sustainable and even preferable from an environmental perspective. Treating the Carter administration's earlier failure as a lesson learned, and informed by Gore's enthusiastic accounts of eco-modernization in other liberal democracies, radical Green-leftists, the pollsters' eco-aware yet deeply individualistic and meritocratic bloc of technocratic suburbanites, as well as blue-collar voters hoping for a return to high employment together lent the eco-modernization strategy legitimacy. This said, the win-win-win solution provided by eco-modernization was to be achieved by abandoning state pretensions to command and control environmental regulation. In place of such policy hubris, the New Democrats' eco-modernization agenda would be anchored by an idea hatched in the same think tanks and sponsored research units that informed the New Right. This was the belief that supply-side management to maximize economic growth will always bring about favorable kinds of development. The Clinton-Gore administration had developed and promoted a policy based on opinion poll–driven perceptions that a significant and effective sector of the voting population were exasperated by the failures of the command and control welfare state economy, and believed that just as society participates in the ecosphere, rather than simply dominates nature, so should government manage rather than master the economy-environment relationship.[11]

By repackaging the jobs-versus-environment debate as a false choice, the New Democrats assimilated the holistic critique of environmental harms to the New Right's critique of collective constraints on individuals' freedom to choose. As had the West European, Japanese, and Oceanic center-left parties, the Clintonite Democrats perceived the discourse of ecological modernization as a political cultural tool. When wielded correctly, eco-modernization could politically empower the New Democrats as it had other third way parties abroad. Optimistically, greening economic growth would reveal the New Right's jobs-versus-environment rhetoric to be a reactionary illusion. Decoupling economic growth from environmental harms in this way, it was hoped, would bring not only lasting electoral success but allow the United States to lead the world in efforts to "save the planet." At a practical level, the eco-modernizing agenda would provoke the moribund American industrial sector into a new round of competitive activity. The resulting free-market induced rising tide of green, techno-efficient innovations would lift all boats. By interpreting once radical green ideas in new ways, the Clinton-Gore administration earned the support of mainstream Greens and a population that opinion research continually suggested was highly sensitive to green issues. The eco-modernizing agenda helped to make the environment an indispensable feature of national and global politics in the early 1990s.[12] The administration's eco-modernization efforts signal the moment at which Radkau's second great global eco-boom had crystallized in the United States.

The eco-modernization agenda made it ideologically possible for one party in a two party system to advocate the wholesale "ecologization of markets and the state." By promoting eco-modernization, the New Democrats had hoped to "shap[e] corporate activity and markets to (re) incorporate environmental externalities into the costs of production" on the basis of what is ultimately a "non-economistic ecological critique" of society's participation in the ecosphere. The New Democrats thus sought to rescue capitalism from itself and, in doing so, vault society into an era of environmentally beneficial "super-modernity" in which market forces guided by rational choices supply socio-environmentally desirable constraints on resource exploitation. Central to the eco-modernization agenda was a holistic and voluntarist "ecological critique [that] requires institutions and ecological considerations to coexist through the prioritization of the latter" without affecting the former. Through the eco-modernization agenda, left neoliberalism emerged as the ideological position of choice for those seeking to "improv[e] the economic and social capabilities of citizens [by administering] a more efficient and productive, cohesive and inclusive, socially and environmentally sustainable, globally competitive eco-state." Eco-modernization takes as given the "naturalness" of free markets within society and, in turn, of society as but one participant in the encompassing ecosphere.[13]

The political authority anchored in the state-market nexus is no longer seen as intrinsically harmful to the environment or citizenly interests and the state itself is no longer legitimately understood as potential deontological regulator of markets and the social relations that obtain within them. Eco-modernization rationalizes the utilitarian calculus in line with the findings of ecological science: the greatest good for the greatest number is recast as a function of economic growth. In keeping with the New Democrats' commitment to a class and race blind prosperity built on equal opportunity in a collaborative green-growth economy, rather than some kind of stigmatizing, bureaucratic nanny state tied to a dirty, conflict-ridden industrial economy, the eco-modernization agenda was conceived as a central support for the universalization of freedom. In short, the New Democrats brought to life as an electoral strategy those modernist dreams of the 1960s and 1970s for an environmentally sustainable, community-based, classless, raceless, genderless, and so benign and tension-free poststate authority. Eco-modernization represents the assimilation to reformist electoral politics of the once radical holistic critique of Enlightenment dualism. The eco-modernization agenda central to left neoliberalism rejects Enlightenment's artificial dualism as an illusion and instead sees everything as connected to everything else. Consolidated in the wake of the first great eco-boom of the 1970s, the concept of the environment had indeed by the 1990s fostered a shift in American beliefs and values. This shift in political culture "challenged the previously hegemonic discourse of industrialism and a nature considered solely as 'resources.'"[14]

Influenced by radicals' ideological and practical successes on environmental issues in the 1970s, the electoral Left could no longer legitimately present itself as representing the aspiration to rational mastery aimed at universalizing freedom. In fact, the first baby boomer presidency had helped to debunk the reenchanting myth of Enlightenment: that of progress towards universal freedom through control of a state dedicated to mastering nature. The third way revolution had undermined longstanding left beliefs that the goal of universalizing freedom as autonomy could be achieved by wresting control of the state and using its institutions to eliminate the social and environmental exploitation wrought by capitalism. The New Democrats cast aside the reenchanting myth of Enlightenment and, in doing so, embraced policies that could have little or no meaning in the traditional sense implied by the modern Left's ideological and philosophical foundations. In the context of a rapidly globalizing postindustrial political economy, the once radical belief that society participates in nature and that individuals bear moral responsibility for ensuring the harmony of this participation provided a newly reenchanting myth of a whole society working in harmony with natural processes. Radkau's "epochal change in human consciousness" had been brought to life, but not on terms of the Left's choosing. As had the New Right in seeking a way out of Hayek's paradox, left neoliberals had given voice to a certain yearning that characterizes modernist political culture, a yearning for a world that once more has meaning in itself, as an ontological feature. In this sense, left neoliberalism grew out of what Geuss, interpreting Adorno, regards as that "deep-seated" change in "the structure of rationality itself" that people on the left had found so difficult to deal with in the mid-twentieth century.

Modernism and the Third Way

Eco-modernization stood in direct and positive contrast with the exhausted discourse of Enlightenment that had sustained the legitimacy of the centralized, liberal-democratic, military-industrial Keynesian welfare state. As Habermas had shown, this legitimation dynamic had depended on the Cold War–framed compromise between big government, business, and labor, which itself had depended on a politically effective, organized modern Left. This modern Left had, it was presumed, represented the generalizable interest of all individuals against a rationalized authority that sought primarily to promote state sovereignty and economic growth. Yet neither eco-modernization nor third way politics depended on such a legitimation dynamic. As James Meadowcroft early recognized, by the 1990s it had become clear to politicians that "no class, movement, or political party could be said to stand in precisely the same relationship to" the much vaunted eco-modernizing state as could be said about the modern Left amidst the compromise that had guaranteed the New Deal and Great Society.[15]

Nor could the eco-modernizing agenda appeal to such a homogeneous social class in the political cultural context of the United States in the

1990s. The solidaristic modern left organizations—such as labor unions and community groups that, not without sectarian and racist biases, once unified and gave voice to the working class and middle America—had dissolved, joined the New Right or been commercialized.[16] The win-win-win argument central to the third way and eco-modernization agenda was not designed to appeal to homogeneous voting blocs arrayed for or against an arrangement to redistribute spoils of social reproduction. Rather, the third way and eco-modernization together held out promises of green economic growth to atomistic individuals, who by their own inner resourcefulness stood to benefit from opportunities to prosper in the new green-growth economy. On the ground, this meant that an idealized classless and raceless citizen would freely choose to start up entrepreneurial ventures and develop environmentally friendly innovations, for example, or to consume with green ethical discrimination, or to retrain for green-tech jobs. Social and environmental exploitation were repackaged as problems of moral willing, that is, as diffuse whole-of-society problems that lack an identifiable class of perpetrators and any identifiable class of the aggrieved.[17]

Buffeted by a recession that exacerbated the large scale offshoring of blue-collar jobs, the Clinton-Gore administration's effort to stifle the jobs-versus-environment debate ultimately failed. Insofar as Clinton, Gore, and boosters such as Thurow and the New Democratic leadership team presumed that eco-modernization would unify party constituencies old and new, and that this support would endure electorally, they were mistaken. Against the backdrop of policy sclerosis and ongoing media attention lavished upon a New Right insurgency within the Republican Party, the New Democrats were comprehensively routed at the 1994 mid-term elections. New Right agitators alleged financial malfeasance, uncovered titillating instances of personal impropriety on the part of Clinton, and undertook blatant red-baiting in relation to the administration's well-publicized yet inept efforts to universalize access to healthcare. The New Right could undermine electoral support for the president and Democrats in Congress by highlighting Clinton's inability or unwillingness to implement policy changes sufficient to revive a moribund economy. In these conditions, the New Right insurgents made significant electoral gains amongst both technocratic suburbanites and the working and lower middle classes.[18]

Tapping veins of real and concocted outrage by stoking the moralized politics of fear, Republicans won fifty-four seats in the House of

Representatives, picked up eight seats in the Senate, and for the first time in fifty years gained control of a majority of state legislatures. Through its widely touted Contract with America, the congressional New Right sought specifically to undo the era of big government, the nanny state, and the White House's environmental regulatory regime. Among other things, congressional Republicans sought a moratorium on the issuance of rules to protect, and demanded stricter cost-benefit studies of, welfare, public health, and environmental regulations and, in a gesture towards those wealthy benefactors sponsoring the Wise Use movement, required government to compensate land owners before they complied with wetlands and endangered species laws. Over the years of "deliberation between environmental and business groups," the Clinton-Gore administration's President's Council on Sustainable Development recommended "over 140 actions. . . . [Yet,] very few of the proposals were even considered" in the face of the ascendant New Right.[19]

In response to the New Right, the New Democrats retreated from even the most modest eco-modernizing reforms. Yet, the Clinton-Gore administration maintained an idealized image of the globalized national polity as a realm of deliberation and consensus seeking rather than of contest to define the distribution of social goods. As such, meliorism was maintained, even as the New Right continued to issue race and class tinged calls to a citizenry threatened by creeping economic and status precarity. Interestingly, the administration's retreat was based largely, once again, on opinion polling that indicated significant support for the quintessential modernist view that all citizens can and should, with minimal assistance, seek to make their own homes within the maelstrom of the globalizing political economy. The president, and those advisors who had not resigned in protest at Clinton's refusal to incorporate social provisions into the North American Free Trade Agreement, interpreted the results of such opinion polls as illustrating what winning votes nationally and in the states required.[20]

Clinton thus took to publicly reiterating the third way view that government assistance should be limited to facilitating equality of opportunities and not outcomes, while also pronouncing a newfound willingness to "get tough on" those unwilling to embrace such opportunities. Meanwhile, the Clinton-Gore administration "put much greater emphasis than previous administrations on expanding the role of environmental information disclosure and voluntary programs."[21] What appears to have helped Clinton to regain some degree of electoral success amidst moral panic over unemployment, crime, and "welfare abuse" fomented by the New Right and encouraged through the newly emergent 24-hour news environment were the Violent Crime Control and Law Enforcement Act of 1994 and the Personal Responsibility and Work Opportunity Reconciliation Act of 1996. The Clinton-Gore team retained the presidency at the 1997 election, even if New Democrats made little headway in the hostile legislature and would continue to lose state legislatures over ensuing decades.[22] For Michelle Alexander, the Clinton strategy was to cynically

> adopt the right-wing narrative that black [and other poor] communities ought to be disciplined with harsh punishment rather than coddled with welfare [by] vowing that he would never permit any Republican to be perceived as tougher on crime than he [and simultaneously] appealing to African Americans by belting out "Lift Every Voice and Sing" in black churches, while at the same time signaling to poor and working-class whites [as well as morally upstanding blacks] that he was willing to be tougher on [recalcitrant] black communities than Republicans had been.[23]

For Democratic Party leaders, what was required was a rightward revision of the original left neoliberal third way policy agenda. Such a revision would maintain the commitment to the provision of "opportunities" in order to garner support from poor white and black Americans, while also appealing to individualistic yet ethically liberal technocratic suburbanites and the revanchist middle strata that had turned to the New Right. The revised strategy was to highlight the moralized politics of fear. Regardless of race or class, some individuals struggling to prosper were trying harder than others and deserved greater access to the "new economy."

In contrast with Alexander's argument, however, Clinton's choral efforts in black churches do not seem entirely cynical.[24] Rather, these were acts born of Clinton's thoroughly modernist conviction that individuals are morally responsible for making and maintaining their own place in an unpredictable, uncontrollable yet manageable world. Clinton's left neoliberalism was grounded in the view that a given quality-of-life outcome could not justifiably be a welfare right attaching to citizenship. Rather, the poor, uneducated, drug-dependent, and wayward, or simply unlucky, were to be seen as bearers of a moral obligation to those who "footed the bill." This moralistic rationale allowed Clinton and the New Democrats to echo New Right accusations that those receiving welfare support deserved economic sanction. Of course, with Alexander, such a rationale did facilitate racist filtering of the "tough on crime" rhetoric. What is important for my argument however is that in "an era of market worship," Clinton had emphasized the moralistic universalist view that "those who couldn't demonstrate self-reliance or independence were [to be] identified not only as unworthy of assistance, but as a potential threat to the core institutions of American society."[25]

The third way thus justified a highly moralized, disciplinary social policy with the reactionary rhetorical strategy of "the reverse Robin Hood," a moralized politics of fear-based strategy that left neoliberal parties the world over would pursue for decades.[26] Those who could or would not demonstrate self-reliance and independence were cast as making immoral choices, as holding ethical orientations antithetical to the universalizing efforts of liberal democracy to promote freedom through self-responsibility and sustainable development through consumer choice. With this move, the New Democrats' post-1994 left neoliberalism repackaged freedom and sustainability as the products of voluntary moral choice. Such a choice was placed at risk by the refusals of young, mostly black and Latino "gang members," the drug dependent, the uneducated, and the unemployed. Similarly, sustainable development was placed at risk by those unwilling to consume with green ethical discrimination. In this way, while left neoliberalism may have served as a racist dog whistle to some, it simultaneously served as an assertion of green liberal universalism to the class- and color-blind—because, demographically speaking, increasingly multicultural and well-educated—technocratic suburban aspirational class. The third way's morally worthy citizen may well be an African-American, Latino, or Appalachian struggling with drug dependency, low levels of education, or under- or unemployment. However, what is important about those struggling under such burdens is something essential about each individual's moral being and not state sponsorship of their access to a part of the spoils of social exploitation of nature. The relevant distinction is an apolitical one made between the morally worthy and the unworthy. What becomes all-important is the moral choice to persevere, even if sometimes falteringly, to aspire and, confronted by mounting scientific evidence, to shop locally and with green ethical discrimination. Such a moralistic, redemptive, ethically aspirational narrative was arguably central to the public appeal of the flawed and all-too-human post-auratic president himself.[27]

Subpolitics and Risk Awareness

The ecological holism and moral voluntarism advocated by radical green-left movements in the 1970s had by the 1990s echoed across the political culture and, in turn, affected how one major party addressed both society and the problem of the environment. Scholars such as John Dryzek, who had critically adopted Habermas's discourse ethical approach to critical theory yet contested his rejection of green modernism noticed something very interesting about this situation.[28] For Dryzek and others such as Brulle, failed or not, the eco-modernization agenda coincided with an upturn in public awareness of the kinds of environmental risks, such as from atmospheric pollution, radiation, or toxics, that the radical Green Left and subsequently mainstream Greens had raised in the 1970s. Dryzek and colleagues critically adapted the concept of risk from work by German sociologist Ulrich Beck. Beck's argument was that increased educational attainment and the emergence of a "cosmopolitan worldly" culture since the 1970s, that is, of modernism, served to generalize public awareness of the inescapable "riskiness of everyday life." Such awareness had once been the preserve of experts but had increasingly been "democratized." Beck identified two factors impacting how widespread awareness of risk transformed the relationship between citizens and political authority. First was awareness of the inherent riskiness of life in modernity itself. For example, along with the benefits derived from automobiles in relation to older modes of individual transportation came the higher risk of a life-altering accident. Second was awareness that the very state-sponsored market institutions that made life in modernity so risky were also unable to control the risks that each produced. Fossil fuels, hydro- and chlorofluorocarbons, atomic energy, and hormone disrupting chemicals delivered benefits. However, the techno-scientific and capitalistic entities supplying such things had little to no control over side effects such as carbon pollution, ozone depletion, nuclear fallout, or hormonal disruption. Less prominent in Beck's work yet unarguably a concomitant was that risk awareness entailed widespread understanding that those controlling such institutions also held an interest in undermining or otherwise slowing efforts to regulate, constrain, or internalize the environmental and social costs of production that markets externalized onto taxpayers.

What is nonetheless important about Beck's thesis is his observation that—just as Meadowcroft had recognized in relation to the politicians' eco-modernization agenda—in risk society mass movements organized to contest capitalism no longer mattered to the legitimation of political authority. Rather, what Habermas had described as a legitimation crisis had by the 1990s exhausted itself. Political authority was by then diffuse and (dis-)organized, legitimated through the interactions of a complex array of subpolitical non-, sub-, and suprastate actors: the Green Left and other new social movements; mainstream green and other nongovernmental organizations; small, medium, and global corporations, their owners, managers, representatives, and advocates; municipal, urban, provincial, and state governments; and international organizations such as the United Nations and the World Trade Organization. Understood in the light cast by Beck's thesis, green-left groups were not only contesting the riskiness of everyday life and focusing on quality-of-life issues while building prefigurative eco-communities but also in many cases were delivering support and credence to mainstream green groups' engagements with techno-scientific and market institutions. Ethically aware—for Beck, "altruistic yet egoistic"—individual citizens, often but not always loosely affiliated with new social movements, were choosing green and ethical products, subscribing to mainstream green and other nongovernmental organizations and charities, engaging in local community and urban politics, and building prefigurative, small-scale alternative, sharing and cooperative economies. The political cultural phenomenon of risk awareness was urging reflexivity on an all but exhausted instrumental political authority. The new risk-subpolitics nexus was ushering in a post- or neo-Enlightenment "reflexive modernity." Fundamentally different from instrumental modernity, this reflexive modernity depended on a multiplicity of consensus driven, bottom-up, quasi-institutional governance forums, each of which exerted only partial and incomplete influence on authority and, in turn, the larger globalization process.[29]

For Dryzek and colleagues, the risk-subpolitics nexus central to what Beck called reflexive modernity heralded a crisis in the state's capacity to legitimate the market political economy. In the risk society,

> nongovernmental actors including social movements solve social problems in innovative ways without relying on the

> administrative state. Subpolitics occurs when greens organize boycotts and protests against corporations—and when they negotiate with corporations to make corporate activities or products less destructive.[30]

What Dryzek and others such as Brulle noticed in relation to Beck's work was that, in the United States and other Western liberal democracies, all of this depended on the success of the left neoliberal weak eco-modernization agenda. More importantly, Dryzek and colleagues exposed how radical green-left modernist groups, although avowedly working sub-politically, depended on state sponsorship of eco-modernization. Often market-friendly supply-side policy-making did serve to dilute radical green-left demands—for greater community control in urban planning or for green civic education, for example—and at best channeled these into narrow voluntarist green consumerism. However, Dryzek and others argued that, by creating space for nonregulatory voluntary initiatives and creating room for negotiation with corporations, even the weakest of eco-modernization agendas might nonetheless open the door to more radical demands. What was taken at the time to be ever-increasing public awareness of risk and concern for the environment might then allow the broader Green movement to rekindle "the association [of the environment] with the legitimation imperative previously seen in the United States around 1970."[31]

In keeping with the spirit of Habermas's metatheory, Dryzek, Brulle, and others saw both the radical Green Left and mainstream Greens as co-contributors to a greening of the public sphere. By bringing green-left concerns to public attention through the media and by acting on locally important quality-of-life issues—by engaging in symbolic actions such as a forest blockade protesting unregulated logging activity, researching toxics in a consumer product, or attempting to influence rural and urban planning and development decisions—the movement as a whole helped to foster "social learning" for an ecologically sustainable, deliberative democratic political system. The argument was that subpolitical actions and communitarian quality-of-life demands, even though not centrally organized or coordinated under the aegis of some hegemonic political project, would over time undermine the systemic imperatives that currently privileged economic growth at any social or environmental cost.[32] The seemingly ever-expanding diversity and multiplicity of broadly green-left modernist demands, themselves seeping into the mainstream, would over time wear down the state and markets, and bring about the desired normative effect of a system conducive to lifeworld expectations. For Brulle, green-left radical and mainstream professional Greens contribute to an "ecological metanarrative" that communicates "the imperatives of the lifeworld to the public sphere . . . serv[ing] as a bridge between the individual's lifeworld and the larger social order."[33] By substantiating a counterfactually open reflexive rather than a closed instrumental modernization process, such an ecological metanarrative would, it was argued, over time replace irrational instrumental systemic imperatives with the ecologically communicative rational imperatives of the lifeworld.[34]

A Force for Freedom and Prosperity

It is through this Habermasian interpretation of Beck's risk society thesis that Dryzek and colleagues analyzed Republican president George W. Bush's avoidance of public confrontation with the Green movement in the early 2000s. For Dryzek and colleagues, now-entrenched citizenly awareness of environmental risks meant that the power of the normative "discourse of environmentalism"

> was learned the hard way by Republicans in the United States, first in their attempt to dismantle environmental standards when they took control of the Congress in 1994 and then in the 2001 development of George W. Bush's energy policy. On both occasions, Republican leaders were hammered by public opinion polls and forced to back off high profile initiatives that contradicted environmental discourse.[35]

Sensitive to an electorate increasingly well informed about pollution, toxic food additives, and quality-of-life concerns such as town planning and urban development, George W. Bush followed both Nixon and his father George H. W. Bush in claiming the middle-ground on environmental and social issues. His administration continued the Clinton-Gore priorities of supporting market-based governance initiatives reliant on moral voluntarism, individual flexibility, and industry consensus. He also promoted New Right ideological beliefs in the importance of citizens' moral worthiness and the role to be played by a national community of patriots in confronting diffuse and ever-present risks. He argued publicly that

> America is a great force for freedom and prosperity. Yet our greatness is not measured in power or luxuries, but by who we are and how we treat one another. So we strive to be a compassionate, decent, hopeful society.
>
> [We are witnessing] a revolution of conscience in which a rising generation is finding that a life of personal responsibility is a life of fulfillment.
>
> By applying the talent and technology of America, this country can dramatically improve our environment, move beyond a petroleum-based economy, and make our dependence on Middle Eastern oil a thing of the past.[36]

As had Nixon in confrontation with the peace and anti-war movement, and George H.W. Bush in response to the backlash against Reaganomics, George W. Bush combined "compassionate conservatism" with "sensible environmentalism." In keeping with the party's longstanding commitment to private over public provision of welfare, his administration enhanced the role played by community-based, charitable, and philanthropic organizations in providing public goods, social cohesion, and upward mobility for the poor and in implementing environmental programs such as urban cleanup campaigns and community renewal programs. Yet in an atmosphere clouded by the terrorist events of September 2001, George W. Bush also authorized a crackdown on the "home-grown terror threat" allegedly posed by radical Green-leftists. All the while, in addition to transferring responsibility for public goods to charitable organizations, George W. Bush used the executive power to weaken environmental protections and, through the Economic Growth and Tax Relief Reconciliation Act of 2001 and the Jobs and Growth Tax Relief Reconciliation Act of 2003, to reduce taxation on the wealthy and so further limit progressive economic redistribution. In this way, George W. Bush worked publicly to unify citizens in a single patriotic community while simultaneously advancing the interests of the Republican Party's "core constituencies, particularly industrial corporations and timber, mining, agriculture and oil interests . . . he drew heavily from those constituencies, as well as conservative ideological groups" to make appointments across a range of agencies.[37]

In light of George W. Bush's public endorsement of consensual horizontal governance, while "working behind the scenes" to advance the interests of "core constituencies" and the New Right ideological agenda, it seems that Habermasian critical theory misses something important about what Beck calls risk awareness and subpolitics. It is self-described post-environmentalist thinkers who offer a more clear-eyed analysis. As did the Habermasians, post-environmentalists also engaged Beck's work on risk society. However, Michael Shellenberger and Ted Nordhaus disagreed with the Habermasian claim that, taken together, radical Green-leftists and mainstream Greens offered a truly rational and so normatively appealing and politically effective counterpoint to the systemic imperative of economic growth at any cost. For the post-environmentalists, as long as mainstream Greens continue to seek technical fact-based solutions to the environmental problem and radicals engage in direct action ecotage and prefigurative eco-communalism, neither would find effective political support. Reformists and radicals alike should appeal directly to citizens' awareness of the risks posed by everyday life. That is, Greens and Green-leftists should appeal not to individuals' ethical beliefs, but rather to material self-interest as citizens being harmed by policy failures. While citizens had by and large accepted once radical arguments that society participates in the ecosphere, and at least a significant sector of the population was ready and willing to make personal moral choices to consume with green ethical discrimination, most citizens simply did not prefer that government prioritize the environment in relation to other policy choices. By emphasizing issues such as the preservation of old growth forests and large ocean mammals, national parks and "paper over plastic" shopping bags, pious Simply Living and organic produce, both radical green-left and mainstream Greens had failed to adequately mobilize public support for the administrative coordination of a strong eco-modernizing agenda.[38]

Meanwhile, public antipathy towards the unquantifiable risky consequences of life in modernity and towards large-scale state and market institutions had increased. As successive Republican and Democratic governments relinquished responsibility for providing public goods, citizenship had become less about rights and responsibilities and more about avoiding all manner of risks while seeking to maintain quality of life amidst the tumult of events. This is the phenomenon that cultural theorists such as Zygmunt Bauman and Richard Sennett at the time described as the quest for a life of meaning amidst creeping "precarity."[39] What the post-environmentalists recognized, and the Habermasians arguably did not, was the broader impact on political authority of the empirical fact that citizens were for the most part aware not only that the riskiness of everyday life was spiraling out of control, and that governments were no longer willing or able to address such riskiness on their behalf, but also that the global corporations responsible for manufacturing many risks were deeply influential on and in many cases had become partners with government in the governance process. The post-environmentalist argument that citizens were indeed keenly aware of environmental issues, but simply did not prioritize them over other socially produced problems, implies that perceptions of risk involve environmental as much as social issues. Climate change, nuclear power, deforestation, species extinctions, food additives, genetic engineering, and toxic chemicals matter just as much as do the prospect of unpredictable terrorist events, urban crime, drug abuse, income insecurity, under- and unemployment, access to healthcare, and educational prospects, for example. Against a backdrop of the shift from political government to subpolitical governance, the normalization of precarity from the 1990s onwards makes environmental risks all but indistinguishable from other risks that a citizen may reasonably expect to be aware of and need to confront as a self-responsible individual "stakeholder citizen."[40]

Such analysis helps to account for the electoral, policy, and political cultural successes of the New Right in the same period. Exploiting the moralized politics of fear to good effect in the wake of ongoing terrorism and large-scale deindustrialization, the New Right offered American citizens opportunities to prosper while undertaking symbolically powerful communal bonding initiatives and patriotic acts of civil privatism. Fulfillment of what George W. Bush had called "a life of personal responsibility" would be achieved by citizens acting to protect themselves, their families, and the national community from exposure to endemic risk. Avoiding the austere economic language associated with Reaganomics by continuing the Clintonian legacy of "retooling" government as the facilitator of opportunities not outcomes, yet rejecting left neoliberal universalism in favor of particularist themes drawn from the New Right's well-funded culture war of the 1990s, George W. Bush conjured an electorally appealing nationalistic and jingoistic popular neocommunitarianism. For political economist Bob Jessop, whom it is worth quoting at length, popular neocommunitarianism

> emphasizes the contribution of the "third sector" and/or the "social economy" (both located between market and state) to economic development and social cohesion, as well as the role of grassroots (or bottom-up) economic and social mobilization in developing and implementing economic strategies. It also emphasizes: the link between economic and community

> development, notably in empowering citizens and community groups; the contribution that greater self-sufficiency can make to reinserting marginalized local economies into the wider economy; and the role of decentralized partnerships that embrace not only the state and business interests but also diverse community organizations and other local stakeholders.[41]

For first-generation popular neocommunitarians, such as George W. Bush, as for left neoliberals, such as Clinton, governing parties should overtly limit themselves to policies designed to offer individuals opportunities to act self-responsibly in their local communities to secure themselves from exposure to the riskiness of everyday life. Indeed, popular neocommunitarianism shares with left neoliberalism the downloading of state responsibility onto individuals and communities, who are called upon to make their own ways amidst a globalization process that is conceived as both natural and irreversible. Also shared is the rejection as illegitimate of conflict-oriented left-leaning political movements, be they radical green-left modernist or labor union. Favored are diffuse, consensus-oriented preference movements that respond to and present new "market solutions" to social and environmental problems. However, popular neocommunitarianism differs from left neoliberalism in one important respect. While left neoliberalism is grounded by an abstract universal—the fully informed, cosmopolitan green calculator of personal and planetary costs and benefits[42]—popular neocommunitarianism is grounded by a concrete particular—the passion-fueled experience of participation in "America." In the 1990s, the third way eco-modernizing agenda channeled green-left radicals' ineffective practical reasoning on the model of structural moralizing into an efficacious justification for authority on the utilitarian model of enactment moralizing. Meanwhile, the first generation of popular neocommunitarianism moved in the opposite direction, softening and humanizing the egoistic calculator in a way that portrayed authority as the instrument of the morally worthy, yet since the 1960s continually wronged, "real Americans."

The Postpolitical Condition

Setting aside this one deep-seated difference between left neoliberalism and popular neocommunitarianism, critical scholarship in the late

1990s and first years of the new millennium began to draw attention to similarities between electoral parties, in the United States and indeed, across the West, that were said to contribute to the normalization of a *postdemocratic* or *postpolitical condition*. These analyses differ in tone and emphasis. However, each is concerned with the hollowing out of what I discuss in chapter 1 as the legitimation dynamic of artificial rationalized political authority. Relying on Williams's political philosophy, I argued that those who hold political power and exert such authority must in the very least appear to be respondents seeking advice on how to administer what he calls the first political question. The critical literature on postpolitics is particularly useful insofar it helps to draw attention to the intentional dissolution or weakening of this dynamic by advocates of left neoliberalism and popular neocommunitarianism. In light of that literature, both left neoliberalism and popular neocommunitarianism appear motivated not by the legitimation dynamic of artificial rationalized political authority but by what Williams calls a political philosophy or theory on the model of enactment moralizing. Indeed, insofar as the parallels that I draw between Williams and Adorno's work holds, the postpolitics literature highlights what Adorno feared would be the overdetermination of rationalized authority, or at least of the identity thinking essential to its functioning.

For political economists and sociologists such as Colin Crouch, left neoliberalism and popular neocommunitarianism both sideline solidary interest-based pressure groups while ushering into power an oligopoly of globalizing corporations and free marketeers. The solidary modern Left had sought to coerce government into "taking on debt to stimulate the economy," so promoting employment and facilitating redistribution of some proportion of the spoils of social reproduction along egalitarian lines. Yet the dissolution of the modern Left in the 1970s freed government to refuse such indebtedness. Such refusal was justified on grounds that financial market deregulation would stimulate employment by encouraging individual citizens to access consumer credit. Such policy is for Crouch little more than "privatized Keynesianism" that shifted the debt burden onto individual citizens while delivering the benefits of economic stimulus to the wealthy and entrenching financial capitalism as the only alternative to economic stagnation. For former third way advocate turned critic Wolfgang Streeck, the increasingly symbiotic relationship between electoral Left and Right combined with the abandonment of politics in favor of eco-communalist withdrawal and ever more radical symbolic politics by what he derides as "the marijuana left" opened the door to a "tax revolt" by the wealthy and aspirational new class. These are the groups that constituted the revanchist middle strata to which left neoliberal and popular neocommunitarian iterations of the moralized politics of fear were designed to appeal. And, in globalization scholar Stephen Gill's view, a truly globalist "new constitutionalism" locks states into a governance regime that is open only insofar as consensus building is dedicated to the advancement of free trade. Within this global governance regime, interests in profitability and rent-seeking trump social interests and constraint on environmental harm, making the default criteria for policy-making of any kind that of market efficiency.[43]

For political theorist Chantal Mouffe, whose response I discuss in detail in chapter 7, postpolitical authority displaces the agonism once associated with state-centered political authority. A complex array of homogeneous, benign, and quasi-formal governance forums has emerged to obscure power differences and conflicting interests and to undermine the modern Left's political struggle to capture state power. In Wendy Brown's view, the emergence of a global governance regime has led to the formation of a distinctive mode of governmentality. As does Luke when recognizing the normalizing of a global environmental rationality or environmentality, Brown argues that the shift to governance has fostered a "neoliberal rationality." This rationality renders individuals' rational choices to self-responsibly maximize human capital the only salient considerations in normative debates over public policy. Once more, Nancy Fraser offers succinct exegesis, finding that these analyses of postpolitics identify a "legitimation deficit," whereby

> democracy's long-standing ills have passed beyond the point of amelioration. . . . [The analysts] link democratic crisis to a mutation in the nature of capitalism. It is the shift from the state-organized capitalism of the [1940s to mid-1970s] era to the globalizing capitalism of the present, these observers claim, that has destabilized political orders throughout the world, hollowing out public powers and turning democratic institutions into empty shells, mere shadows of their former selves.[44]

What the postpolitical commentators' recognize is that this "hollowing out" of "public powers"—which for better and worse the modern Left had addressed as powers centralized in the state—has not only undermined citizenly confidence in state institutions but also citizens' ability to contest authority. Citizenly confidence that the state can effectively order social relations and reproductive processes in the generalizable interest has been dissolved. Simultaneously, even when citizens do contest a particular set of institutional arrangements, the results are more than likely to be ineffective or inconsequential. In these two senses, the postpolitical condition represents an epochal shift. The shift from "government to governance" or "organized" to "disorganized capitalism" has altered the legitimation dynamic that had obtained until the mid-1970s. The dissolution of the modern Left that had not unproblematically represented citizens' generalizable interests has left only the business lobby representing market interests in profitability and rents, and those holding power in the state.

Since the 1990s, a pseudo-legitimation dynamic has been shaped by two notable shifts in the way that the electoral parties work to obtain and maintain power. As Habermas saw, parties seeking government by electoral means no longer need to appease an organized bloc of citizens pretending to act in the generalizable interest. However, parties are not entirely free to favor market interests without some legitimating justification. As Ingolfur Blühdorn suggests, electoral politics and policy in a "postecological" holistic social order would no longer need to respond to pressure from a citizenry organized into mass movements but, rather, to quantified representations of individual preferences or opinions in relation to select "issue frames": the environment, jobs, drugs, terror, crime, abortion, or immigration, for example. Legitimation has in his view come to center on parties' analyses and responses to select voters' "revealed preferences" for policy. Parties aggregate the preferences expressed by samples of voters.[45]

These economically rationalized issue frames also quantify social movement demands, much in the way that firms are alleged to respond to price signals from consumers. Such demands are conceived as factors used in the design of policy to appeal to discrete blocs of voters. In government, parties enact such policy in accordance with the principle of economic efficiency. The risk-aware and precarity-exposed individual opinions to which the data give voice in this way subsume overt appeals to ideological vision. Party analysis of such data sets stands in as a proxy for what were once ideologically laden interpretations of public opinion. For Democrats, this was a progressive left-liberal interpretation of the generalizable interest prevalent until the late 1970s. For Republicans, such interpretation supported the moderating voice of patrician conservatism, which the Goldwater insurgency first challenged in the early 1960s. The shift from an electoral party as the ideological interpreter of public opinion into the positive analyzer of aggregated data on voter preferences not only reflects the market triumphalism that began with the collapse of the Soviet Bloc in 1989 but also the influence of well-funded efforts to bring the free-market philosophy and the tools of rational economics to bear on political problems. Primarily influential was, of course, the peculiar combination of holistic naturalism with moral voluntarism under the aegis of Hayek's thought, which I discussed in chapter 3. As recently shown by Sonia Amadae and Nancy Maclean, also influential was the Virginia school's iteration of the public choice doctrine and, as will be discussed in detail in the next chapter, the now predominant form of jurisprudence associated with the Chicago school of law and economics.[45]

By pretheoretically defining voters as rational choosers oriented to maximizing utility, party and, in turn, governmental focus is oriented toward those who have already made the rational choices necessary to flourish as individuals. In such a climate, the legitimacy of political authority depends on the capacity of those holding government to avoid provoking only those with something measurable to lose, such as Streeck's tax revolutionaries, and the revanchist middle strata of precariously situated suburbanites at risk of "falling through the cracks." In this perspective, postpolitical governance offers the lucky few uninhibited access to a high quality of life while, through the personal credit industry, offers market interests a multitude of commercial opportunities that need not be "made" but only "purchased in America." Simultaneously, citizenly demands for an egalitarian distribution of exposure to pollution or income, for example, appear as irrational demands for some form of public welfare incommensurate with economic rationality. A movement claiming that pollution unjustly affects the poor or demanding a living wage or welfare appear either as champions of the undeserving or dangerously disruptive. Such movements are open to mass-media portrayal as hyper-radicals seeking to ignite race or class war.

In a society polarized between risk-aware and altruistic cosmopolitans and egoistic utilitarian rational calculators, exploitation is filtered through the lens of subjective preferences for opportunities to escape pollution and poverty or to relocate to locales in which organics are more readily available and one might telecommute. From the electoral perspective, issue frames allow political actors to highlight the "pressure points" around which already predisposed voter blocs—disaffected working class whites or entrepreneurial migrants working in the infotech industry, for example—might be outraged. In effect, parties become postdemocratic organizations engaged in postpolitical competition over starkly opposed moral visions.[46] Such visions portray authority as justifiably embodying the benevolent force of a green global invisible hand or the red in tooth and claw Darwinian forces of natural selection. From the late 1990s onwards, risk-aware and precariously situated citizens' "incredulity"[47] at the myth of progress toward freedom through rational mastery would morph into "discontent" with left neoliberal offers of equal opportunity to all. Meanwhile Republican Party advocates of popular neocommunitarianism, even as they advanced technocratic and market-friendly neoliberal reforms, would hone their appeals to real Americans and provoke calls for a Tea Party-like revolution to rekindle national greatness.

Chapter 5

Postpolitics and the Return of Moral Authority

The Externalization Thesis

In the mid-1990s, arguably for the first time since the early 1970s, radical activism inspired by green-left modernist ideas gained significant purchase on public affairs, both within the United States and globally. Motivating this radical revival was awareness that, while gesturing towards progressive environmental and social reforms, the third way and Washington Consensus had loosened constraints on private property rights and prioritized wealth maximization as a social goal. In the wake of the collapse of the Soviet Bloc, such neoliberal reforms had been presented as liberatory efforts to end dependency, to promote opportunity, and, of course, to "green" economic growth. However, in practice the shift from welfare to workfare and from command and control regulation to horizontal governance was providing tax relief and possibilities for endless negotiation over social and environmental protections to private businesses and corporations. Meanwhile, the third way had brought into being a punitive "market discipline" that was applied to the poor and minorities, had created a burgeoning un- or under-employed "educated precariat" while marketizing public and higher education, and, at the level of popular culture, facilitated the demonization of the poor as unwilling to consume with ethical discernment or aspire to a better quality of life.

As anointed in homage to the radicalism of the 1960s by news media, this new New Left opposed the deregulation of markets and regressive taxation policies that both electoral parties had facilitated since the 1980s. Effectively flipping the efforts of public choice theorists to use game theory to elaborate a comprehensive account of why the state should constrain collective actors in order to facilitate economic growth, scholars and activists associated with the new New Left argued that states, through the undoing of social and environmental regulation, were unjustly allowing market actors, especially large global corporations, to benefit from the production and maldistribution of social and environmental harms.[1] A key focal point for public protest was the view that politicians and policymakers, alongside lobbyists and legal advocates, worked to sustain rules that unjustifiably allowed market actors to externalize the social and environmental costs associated with profit-seeking. Such rules left citizens in general, and more often than not minorities and the least well off in particular, ultimately responsible for bearing the harms produced. What the protesters made clear was that a large proportion of the benefits accruing to business as a result of externalization were, in fact, illegitimate. The ostensibly radicalized externalization thesis highlights the fact that businesses extract gains from society and nature that could be factored into the costs of production yet, for political reasons, are not. With government support, firms externalize these factor costs onto society as a whole, and accumulate the savings as "free money," what all strands of economic thought from Marxism to social choice theory decry as rents. What the flipped externalization thesis popularized in the 1990s were views that profitable market actors should meet the costs of cleaning up pollution, caring for those harmed by it, and the costs associated with welfare support to those paid low wages.[2]

This iteration of the externalization thesis gained significant public support insofar as, for over a decade, the consequences of market-friendly policy-making had received considerable news media attention. The most publicized issues, several of which Beck and his anglophone interlocutors discuss in detail as contributing factors to the emergence of a global risk society, were perhaps the Love Canal disaster and cover-up in New York State in the late 1970s; the nuclear meltdown at the Three Mile Island reactor near Harrisburg, Pennsylvania in 1979; efforts to dump toxic contaminated waste in predominately black communities, such as in Warren County, North Carolina, in the early 1980s; the 1984 chemical spill at the Union Carbide Plant in Bhopal, India; the 1986 nuclear disaster at Chernobyl in the then USSR; the 1989 grounding of the oil tanker Exxon Valdez in Alaska; Shell Oil's attempt to sink the oil platform

Brent Spar in the North Sea in 1995; and the systemic exploitation of low-cost foreign labor and lax industrial and environmental regulation in the global South by highly visible corporate brands such as Nike, Reebok, J. Crew, Calvin Klein, Gap, Starbucks, and Apple.[3]

The "commonsense" or ah-ha factor arguably attaching to the externalization thesis prompted a powerful reaction from the business community, which is the main focus of discussion in this chapter. In chapter 6 I return to the radical movements of the mid-to-late 1990s and discuss the ideas and scholarship informing this new wave of green-left modernism. The radicalization of the externalization thesis prompted many in the business community and, subsequently, major electoral parties to resuscitate longstanding yet, until the late 1990s, relatively inconsequential free-market ideas of corporate social and environmental responsibility. Corporate responsibility rebrands corporate owners and managers seeking profit as enlightened reformists confronting a moral choice to do noneconomic good by doing well economically. Encouragement for the corporate responsibility movement came from university-based business management schools seeking to enlist students who, while perceptibly sympathetic to free market ideals, also opposed the exploitation that the externalization thesis bought into view. One permanent and striking feature of advocacy for corporate social and environmental responsibility is that those promoting and practicing it did not dismiss out of hand the injustices highlighted by radicals. Advocates of the corporate responsibility countercritique instead promoted a "market for virtue" in which responsible corporations would, if left unfettered by harmful regulation, displace irresponsible corporations and so bring about a benign or even beneficial "natural capitalism" or "compassionate capitalism."[4]

Whereas during the Reagan revolution, Milton Friedman's claim that "the sole social responsibility business is to increase profits" had sufficed as a riposte to both the radical and reformist Left, free-market think tanks, funded and unfunded university centers and institutes, business lobbyists, and, importantly, socially engaged and eco-aware entrepreneurs with roots in the 1960s and 1970s counterculture who supported corporate responsibility would avoid such dry justifications for profit-seeking. Indeed, the ideas and practices of the corporate social and environmental responsibility countermovement were portrayed in epochal terms. The voluntary adoption of such responsibility by corporate owners and managers was represented as evidence of a paradigm shift encompassing capitalism in its entirety. Many advocates and practitioners cast the responsible limited-liability corporation, its owners and managers, as avatars of a genuinely holistic critique of an outdated dualism that had falsely sustained an "enemy perception," which unhelpfully divided citizens from markets.[5] Such advocates were sternly critical of the failed dualisms of industrial society and called for holistic and integrated approaches to sustainable development, for more holistic thinking and decision-making in business, and for a holistic view or an approach to doing business that can ensure problems are addressed holistically.[6]

Unsurprisingly, the countermovement received a great deal of support from presidents and politicians from both major parties. When taking office in 2008 and almost immediately called upon to regulate in response to the Great Recession, Democratic president Obama emphasized the responsibility of business to American society. Following George H. W. Bush, Clinton, and George W. Bush, Obama rejected out of hand the need for external deontological regulation of markets on grounds that moral voluntarist measures would suffice to constrain market excesses. As had these predecessors, and Reagan and Carter before them, Obama eschewed calls for regulation as outdated. Rather, in the wake of his well-publicized "green inauguration," Obama reassured the country that given the largesse bestowed upon them by citizens in order to rescue the situation, most corporate owners and managers, bankers and financiers would willingly shoulder their obligations. As postpolitics scholar Brown notes, however, similar to George W. Bush's response to the terrorist attacks in September 2001, Obama did not emphasize the political responsibility of government to constrain but rather emphasized all citizens' shared obligations to contribute to a sustainable and secure return to economic growth.[7]

In this sense, Obama's call echoes the Hayekian holistic critique that seeks to limit authority's remit over the exercise of power in society to that of *catallactics*. To recap, entailed in this view is the belief that political regulation of markets will always and everywhere produce greater negative consequences than would the aggregated choices of moral volunteers acting in markets. Whether announced by presidents responding to terrorism or financial crisis, or by business "thought leaders" in response to the externalization thesis, the interlinked ideas of corporate responsibility, good corporate citizenship, and consensus-driven governance together help to delegitimize views that a deontological political authority, which implements regulation based on principles different from and potentially opposed to those operating in markets, could be beneficial to society as a whole. Taken up by both electoral parties and by the time of the financial crisis representing the corporate mainstream, the idea that corporations, their owners, and managers, bear ultimate responsibility for internalizing externalities helped to undermine justifications for the use of regulatory means to account for the negative socio-environmental consequences of profit-seeking. The combination of holistic ecological thinking with the prioritization of moral voluntarism allowed those in positions of authority to dismiss out of hand possibilities that deontological institutions may provide for better social or environmental outcomes than would the aggregate sum of choices made by ethical over unethical moral volunteers. Advocacy for responsible capitalism simultaneously allowed many of the major corporations targeted by radicals to recalibrate their public images in order to better embrace the all-important market segment of well-educated, articulate, and ethically eco-aware consumers, shareholders, and other community stakeholders prioritized under the turn to governance.[8]

Concomitantly, the corporate responsibility countercritique undermines that dimension of the externalization argument positing that a corporation's status as an artificially constituted interest collective implies that its license to operate could and should refer to nonmarket principles. Portraying government as inefficient, incompetent, and anyway powerless, advocates for the corporate social and environmental responsibility movement claim that "the obligation of constructing new regulations" should be left up to moral actors themselves, be they corporate managers, individual consumers, or the local stakeholder communities in which they operate.[9] In this way, the movement also follows the longstanding right ideological tradition of portraying appeals to political authority—in this case, employment of the externalization thesis as a justification for regulation according to nonmarket principles—as utopian idealism supported only by irrational radicals or overzealous and anyway inept but well-meaning politicians. The modern idea that the corporation is the artificial concessionaire, appointee, construct, or agent of the state is replaced by one that regards the corporation as ultimately the natural manifestation of free moral agency within the encompassing whole ecosphere. The corporation in this view has been selected for under the Hayekian rule of (common) law in order to enhance humanity's capacity for making the moral choices that will desirably influence humanity's participation in the ecosphere.

The corporate responsibility movement is in this sense a political movement with a depoliticizing aim. The core ideological claim of this countermovement's critique of externalization is in fact the moralistic view that political authority evoking principles external to those constituted through free exchange will always create a moral hazard, which undermines the superior efficiency of markets. Rules derived from nonmarket principles—for example, to protect the riparian border of a waterway as a public park or safeguard a living wage to uphold fair working conditions—weaken moral actors' natural resolve to do good by doing well. Regulatory approaches that are insensitive to the market dynamic thus distort the normatively desirable competitive equilibrium that markets produce: actors will always attempt to circumvent or exploit the park's boundaries, and will only pay the bare minimum wage and duty of care to employees. In critic Ronen Shamir's view, the countermovement argument is that, "mandatory rules create a race to the lowest common denominator and a corporate culture of 'box-ticking.'"[9] For Shamir, as well as for Eve Chiapello, the countercritique shifts blame for the production of socio-environmental harms to artificial political authority itself, for propagating a morally lax attitude of minimal compliance. With this move, the corporate responsibility countermovement prioritizes freely choosing moral actors who, as good corporate citizens, attune themselves to the needs of stakeholder communities, the demands of ethically aware consumers, the national interest and ultimately the planet.[10] Shamir argues that advocates of corporate social and environmental responsibility actively urge policymakers to

> go beyond the 1980s politics of deregulation on the one hand, yet also accepting the failures of state-centered command-and-control regulation on the other hand, the authors advocated a third way that was essentially identical to that of governance: greater participation of nonstate entities in the regulatory process and greater emphasis on dialogue and persuasion rather than sanctions and adversarial methods as a means to ensure compliance.[11]

As do left neoliberals, advocates of corporate social and environmental responsibility endorse the so-called shift from government to governance as a step toward the universalization of freedom as autonomy. Moral actors should not in this view be tempted to create moral hazards by supporting regulation. Rather, actors should voluntarily avail themselves of opportunities to produce and consume with ethical discrimination, such that the production and maldistribution of environmental and social harms will over time be reduced. Just as the Clinton-Gore administration had tacitly recognized that American workers were being harmed by policies designed to promote competition but neutralized these criticisms by laying blame for such harms on morally deficient individuals, this market-based third way maintains the overall normative thrust of the externalization thesis but neutralizes the rationale for regulatory demands by moralizing the production of social and environmental harms. However, as do first-wave popular neocommunitarians such as George W. Bush, the movement also appeals to nationalist and patriotic sentiments. The movement seeks to rescue free-market capitalism and free-enterprise people from the regulatory threat. Corporate responsibility does this by linking the moral voluntarism exercised by empowered individuals and communities to both post-1960s ethical holism and the "traditional" American civil privatism that rejects the politicization of allegedly private choices.

The countercritique in this way redefines corporate irresponsibility as a moral quandary confronted by business owners and managers and the individual and community stakeholders with whom they deal, be they former 1960s West Coast countercultural figures such as *Whole Earth Catalog* founder Stewart Brand or heartland entrepreneurs such as the Walton family, owners of prominent practitioner of corporate responsibility, Walmart. The countercritique is depoliticizing because it obscures understandings of irresponsibility as a political problem of the structural constraints that facilitate or enable—license—profit-seeking enterprise. The countermovement repackages profit-seeking, uncoerced by political authority, as a noble task undertaken by the morally worthy who see the economic value of good choices to boost rather than undermine socio-environmental goods. The movement positions business and consumers, albeit via their revealed preferences for ethical goods and services, as the most appropriate agents for addressing questions relating to the production, distribution, and amelioration of social and environmental harms. And it portrays political actors who reject even the radicalized externalization thesis as insufficiently critical of market-friendly policymaking as opponents of individual, community, and environmental self-improvement. Embraced by popular neocommunitarian and left neoliberal electoral parties and, indeed, by many radical Green-leftists, the discourse of corporate social and environmental responsibility repackages the political critique of injustice as a morality play. The externalization thesis leverages the once radical holistic critique of Enlightenment dualism as a commercial asset. The corporate social and environmental responsibility movement recuperates green-left modernist expectations that in the 1970s had sought a system in which life would be meaningful, authentic, and grounded in ethical self-awareness while conducted in harmony with nature and, indeed mimetically express benign natural values.

Communicative Rationality in the Age of Ecology

It is Blühdorn who most keenly recognizes the problematic political ramifications of 1960's and 1970's radicalism. In Blühdorn's work, the once radical holistic critique of Enlightenment dualism and its concomitant prioritization of moral voluntarism fills the institutional void left by the collapse of the military-industrial Cold War economy, the welfare state, and the solidary citizen movements that helped to sustain it. In the so-called risk society that emerged in the 1990s, radical holism and moral voluntarism inadequately fulfill roles once played by deeply politicized institutions. What the subpolitics of risk society brings with it are genuinely more deliberatively, discursively, and so ecological-communicatively rational kinds of democratic and quasi-democratic institutions, the most prominent of which being stakeholder- and community-engaged responsible corporations. For Blühdorn, the new quasi institutions of deliberative local-global governance are

> a form of political agenda-setting, decision-making and policy-implementation that fuses elements of traditional-style, state-centred *government* with the [radical] social movements' *new politics* ideal of autonomous, grass-roots democratic self-rule.

The moral imperative of contributing to the establishment of an ethical global sustainability or, alternately, a national communal greatness, have emerged as key factors of authority's capacity to stabilize the political economy in an extremely risky world. Radical green-left modernist ideas that in the 1970s successfully challenged the Enlightenment myth of progress through rational mastery, by the first decade of the new millennium had come to justify not taking the action necessary to prevent system collapse. Radical green-left movements once saw decentralizing power through the participation of average citizens and local communities in deciding political affairs as the means for undoing the pathologies of Enlightenment and its falsifying myth of rational mastery oriented to freedom. For Blühdorn, it has become authority's efforts to fulfill the very demands of 1960s and 1970s radical Greens and leftists that currently serve to legitimate the administration of an unsustainable system. Formerly radical objectives have come to justify a governance regime premised on a citizenry of self-actualizing consumers, active in their local communities yet eschewing politics proper.[12]

The point to take from Blühdorn's analysis is not that the turn to governance and corporate responsibility are shams, but rather that advocates of both do respond to ethical individuals' subpolitical moral choices. Rather, the point is that risk-aware citizens do not organize collectively to challenge the system by exposing the structural conditions that facilitate externalization. Rather, citizens seek individually and voluntarily to realize themselves as ethical contributors to a wholly ethical society, be it one envisaged by left neoliberals as operating on the global scale and in harmony with benign nature or as envisaged by first-generation popular neocommunitarians, as the embodiment of a God-fearing cosmological struggle against evil. That is, the quasi institutions of the turn to subpolitical governance—the government and politicians that listen in "Town Hall" meetings, the corporations who engage with community stakeholders, the risk-sensitive, eco-aware consumer boycotters—have operationalized central motifs of the once-radical holistic critique of Enlightenment dualism. While radical in the 1970s, decades later subpolitical concerns with individual self-sufficiency, community empowerment, and proximate ethics of trust had come to constitute the basic building blocks of the domestic and global governance regime. The pretense of those responsible for managing a corporation or governing the nation is not that of achieving ever more rational mastery but that of achieving society's harmonious participation within the whole ecosphere or the Christian God's cosmological order. Ideologically speaking, left neoliberals seek the universalization of opportunities for actors to make eco-aware and so rational moral choices in a truly global and totalizing market founded on the benign ontological truth that everything is connected to everything else. Such left neoliberal iterations of postpolitical moral authority privilege responsible corporate owners and managers, policy advocates of a green New Deal and worldly fair trade consumers. As would become clear during and after Obama's second term, such neo-Enlightenment holism and moral voluntarism could also well sustain a second-generation popular neocommunitarianism that, dispensing with the need for artificial constraints on the exercise of power by the righteous, would privilege true patriots who reject cosmopolitan niceties and focus the competitive spirit on greatness and the regeneration of forgotten heartland communities.

In the context of what Radkau describes as a neo-Enlightenment age of ecology, the emergence of which is to a considerable degree traceable to the first great "eco-boom" that began in the late 1960s and early 1970s, the Habermasian concept of a counterfactual communicative rationality grounded in the inherently just discourse ethics of the lifeworld becomes problematic. When understood in historical terms, rather than as a counterfactual abstraction, the communicative rationality that Habermas identified with the ethos if not the practice or arguments of green-left modernist movements in the 1970s and 1980s and later in the 1990s with the stabilizing force of the law that expresses universal principles seems to emerge as a positive justification for the status quo. Even Dryzek's efforts to "rescue communicative rationality from Habermas" with an ecologically communicative rationality that would "underwrite respect for natural objects and ecological processes" seem belied by the real conditions that accompany the subpolitics of risk, the governance turn, and the rise of corporate responsibility. To paraphrase well-rehearsed critiques of the Habermasian idea, when brought into contact with power, as the counterfactual must be if it is to influence authority, the negativity of the uncoercive forceless force of the better argument seems all too readily amenable to coercive positive fact.[13]

In theoretical terms, when operationalized in historical and social context, the structural moralizing commitment to practical reason grounded by inalienable truth is tarnished by reality. That reality is, in the historical United States, one in which the enactment moralizing commitment to practical reason grounded by the egoistic utilitarian calculus is prioritized. The communicative rationality of the ideal speech situation certainly expresses a real condition of human social interaction. But discursively redeemable rational argument and political contest are two different categories of human experience. Acts are universally justifiable if all affected parties would agree on the basis of an open discourse. But the closest practical approximation of such universal justification, political economic relations guided by the subpolitical principles of governance for sustainability, or indeed, greatness, seem to sustain outcomes that critics of injustice as unfreedom otherwise find oppressive, exploitative, or at the least the normatively problematic perpetuation of unsustainable or unfair practices.

It seems fair to say that this is a consequence of rational ecological communication's ready enlistment in the service of best-practice oriented market agency, of Blühdorn's system of unsustainability. Following decades of left neoliberal and popular neocommunitarian incumbency, the once radical green-left demands most amenable to market interests—holistic ecology, be it benign or malignant in conception, and the ethos of moral voluntarism inherent in interpersonal communicative settings[14]—have emerged as key contributions to the exercise of a postpolitical moral authority. The hitherto radical green-left and communitarian critique of the system, posed from the standpoint of the moral volunteer in proximate communities and so holding access to the inalienable truths of holistic ecology and an ethical lifeworld, has been redeployed as an expression of a logic identical to that on which the system has come to operate. The engine for this construal of the self-directing individual able to successfully inhabit autonomous communities and so ensure their harmonious integration is the very political cultural modernism that Habermas had eschewed in his opposition to the Nietzschean aporia. *Modernism* was the basis for critique of *modern* political culture turned pathological in the mid-twentieth century. The problem is that the urge to autonomously make one's own home in modernity—to oppose nonidentity against identity thinking—had by the late twentieth and early twenty-first century itself emerged as the pathological justification for untrammeled primitive freedom, false nonidentity thinking.

Like Blühdorn, Luke is also more skeptical than post-Habermasians such as Brulle, Dryzek, and colleagues about the emancipatory possibilities opened by the turn to governance, the subpolitics of risk and the rhetorical and practical successes of the corporate social and environmental responsibility countermovement. For Luke, it is the moralizing tendencies of once radical green-left modernist values that should be the focus of critique, insofar as these depoliticize injustices such as the maldistribution of pollution, quality of life, or the effects of climate change. As seen, from the late 1970s onwards Luke has argued that the transformation from a production to postindustrial political economy would render many hitherto radical green-left modernist ideas expressions of artificial negativity. Holistic ecology and the moral voluntarism underpinning ethical eco- and risk-aware choices offer a soothing and appeasing myth. This new myth fulfills more deeply seated post-Christian yearnings for escape from the antagonism, underhanded brinkmanship, and deceitfulness of politics. The spread of holistic ecology and ethical eco-awareness through the governance turn offers frustrated progressive suburbanites and the new infotech class a high moral ground from which to participate in the development of a friction-free, environmentally sustainable universe of ethically acceptable consumption and production. Observing this transformation of political culture in the context of a changing political economy, Luke traces the "origins and operations" of corporate social and environmental responsibility programs to the peculiar form of "green governmentality" that accompanied the rise of left neoliberalism and the governance turn in the United States.[15]

Luke finds that the turn to governance signals a shift in the balance of interests required to maintain the legitimacy of authority *as* moral authority, as a governing mentality. Holism and moral voluntarism offer political rulers and the ruled, sovereigns and subjects, a universal grammar that affords potentially all citizens with opportunities to appear to each other to be acting rationally, to be thinking and doing good. The combined effect of decades of discursive and policy interventions prompted by the radical Green Left and, subsequently, mainstream Greens, the scientific community, and politicians eager to please a significant fraction of well-educated and articulate voters, as well as businesses eager to supply such voters qua consumers with environmentally friendly products, has been the dawning of a governmental regime premised on the naturalistic fallacy. At once material and ideological, such eco-governmentality produces rational actors and imposes an ecological rationale for political decision-making that readily admits, and indeed requires, artificial negativity.[16] What Luke calls the "system of sustainable degradation permit[s] those in power to draw upon capitalist culture itself in order to say 'something is being done.'"[17] As such, by

> conceding that earlier industrial market thinking misunderstood the coevolving needs of the economy and ecology by ignoring the promise of natural capitalism, businesses' CSR units now can work plans based on biomimicry analysis, industrial metabolism or c2c (cradle-to-cradle) design to harness sustainability.[18]

It is not by contesting externalization as a political problem but rather by making well-informed and well-intentioned good green choices that all actors—from individual consumers to global megacorporations—might prove themselves worthy. That is, it is not by organizing collectively to coerce political authority into constraining or disrupting the accumulation of profit but by aligning subjective interests with the demands that sustainable development or indeed freedom and prosperity places on society as a whole, be it conceived in cosmopolitan globalist or communitarian nativist terms. In this view, the governance turn implies that neither responsible firms nor green-growth oriented policymakers seek to "reflect nature in social practices," nor do they work to "derive social practices from nature." Rather, such actors resolve the problems of maintaining power and preserving authority by being nature, by expressing, replicating, copying, and, ultimately, *imitating* nature. Most telling for my argument is Luke's observation that with the turn from government to governance, something very interesting takes place. Sometime in the first decade of the new millenium, "biomimicry reasoning comes full circle."[19]

That is to say, prevailing justifications for authority, be they left neoliberal or popular neocommunitarian in tone, have revived nothing short of that which Horkheimer and Adorno saw as cardinal in traditional societies without or prior to Enlightenment: the moral authority of a power that expresses nature, mimesis. What Horkheimer and Adorno in the 1940s viewed as the traditional pre-Enlightenment myth of reenchantment, the mimetic relationship with nature—the imitation, copying, replicating, or expressing of nature[20]—literally and figuratively returns. Mythmaking justifications for authority once more have come to anchor themselves as expressions of a universe having meaning in itself, as an ontological feature. With the erosion of the self-evidence of the modern Enlightenment dualistic worldview, politicians and policymakers upholding consensual governance, responsible corporations, and their owners and managers, came to serve in the same roles as do those who, for Horkheimer and Adorno, hold together the cosmological order in premodern society. Artificial political authority's pretensions to measurement, predictability, order, and control of nature for human ends are relinquished. In their place are established competing naturalistic visions for a holistic moral authority. In terms of authority's legitimacy, public assessment of the mimetic relationship's quality once more provides the basis on which elites justify their authority. Authority and with it meaning itself are once more decentered, and anchored in the cosmological order. In mythical terms, it is once more elites' capacity to ensure that social relations express the natural order and not to manipulate that order which guarantees human flourishing.

Postpolitical Moral Authority and "Neoliberal Jurisprudence"

At the dawn of what Radkau calls the age of ecology, the once radical holistic critique of Enlightenment dualism had come to serve as neo-Enlightenment justification for the displacement of artificial political with moral authority. As Adorno had foreseen in the mid-to-late twentieth century, when authority justifies itself as the expression of the natural order, the will of the most willing is placed beyond negative critique. Negative critique is rendered heretic and protest becomes subversion of the true order of things. Holistic, nature-expressive mimetic authority refers not to rational proofs, to evidence, but to the immutable facticity of nature. Holistic moral authority refers to some benign conception of nature as a harmonious haven that is just when balanced in equilibrium. However, as Adorno feared, when holistic moral authority refers to some malignant conception of nature as a competitive struggle in which survivors demonstrate their higher worth, positive agency is reduced to blind calculation. Highly important in this regard have been two relatively recent developments in law and jurisprudence, one ostensibly left neoliberal in tone, the effort to revive the nineteenth-century doctrine of legal pluralism, the other a direct product of the New Right reaction to the crisis of the 1970s, the well-funded effort to make the theory of law and economics the prevailing orthodoxy.

In one area of the law, a revival of the late nineteenth-century liberal doctrine of legal pluralism drew inspiration from the postcolonial cohabitation of modern secular with traditional law under globalization. For contemporary legal pluralists, such cohabitation serves as empirical evidence for the falsity of the Enlightenment "modern assumption that all legal and political theories can be reduced to the relationships between Sovereign and Subject or State and Individual." In the pluralist view, the governance turn liberates groups of free private persons to "exercise within the area of their competence an authority so effective as to justify labeling it a sovereign authority." Therefore, it is argued, "government must recognize that it is not the sole possessor of sovereignty and that private groups within the community are entitled to lead their own free lives."[21] Extending such observations to address the rise of corporate responsibility, legal pluralists seek to reorder the state-subject relationship along new lines. The contemporary legal pluralism simultaneously reduces the rationalized authority that is institutionalized in the state apparatus while magnifying the status of free private persons as lawmakers within legally defined areas of competence. In effect, the regulation of private activity, such as resource extraction and energy production or industrial processing, is handed over to the private property owners whose actions impact the environment or workplace health and safety.

In another arguably more influential area of the law, advocates of the law and economics movement also lend support to the corporate social and environmental responsibility movement. Gaining ground in law schools and on benches across the United States since the late 1970s, law and economics is a product of the decades-old efforts of scholars and activists to undo the crisis of democracy. It seems unarguably to be the case that, like public choice theory in the social sciences, the law and economics movement was explicitly funded and designed to aid such efforts in the spheres of legal scholarship, legal education, and, most importantly, jurisprudence. Seeking to "bring a corporate-oriented cost-benefit analysis to regulation and legal liberalism more generally," and to "undermine the [allegedly leftist] intellectual foundations" of modern jurisprudence, advocates of law and economics seek to facilitate "'public interest' lawsuits on behalf of property owners." Law and economics helps therefore to consolidate the broader New Right aim of liberating those holding private property rights and so, contributing disproportionately more to economic growth, from legal interpretations of the public interest that might prioritize nonmarket principles and therefore impede or slow such growth.[22]

Insofar as these two developments in legal theory and jurisprudence support the governance turn and corporate responsibility, legal pluralism and law and economics represent two sides of the same holistic, methodologically individualistic, and thoroughly modernist coin. In this sense, I follow Alain Supiot and address the two together as defining a single trend, which he derides as the expression of a panoptical "mania for quantification" and that the postpolitics literature at times rather loosely critiques as neoliberal jurisprudence. These two developments in the law represent the tacit rejection by legal practitioners of the modern Enlightenment view that the law primarily embodies the contingent interpretation of rationalized authority's legitimate response to Williams's first political question. Each rejects rationalized authority's artificial status as the product of contest amongst power blocs representing different and not necessarily market-based or derivable interests, some of which the law may justifiably defend against markets. These developments in legal theory and jurisprudence, then, signal a transformation in beliefs about the law.[23] Insofar as defending private sovereignty, property rights, and wealth maximization are the only justifiable options for legal decision making, law emerges as the depoliticized vehicle for enforcing practical rationality on the utilitarian model of enactment moralizing. Dispensed with are understandings of the law as the historically sensitive yet principle-based exercise of judgment in relation to contest among interested parties and oriented to some settlement that defines public welfare.

As does Hayek's economistic philosophy, law and economics in particular similarly revives a premodern ideal. In the case of the latter this premodernism takes shape as a doctrinal purism in which jurisprudence appeals to and defends the pure truth of Darwinian struggle as staged in a technologically sophisticated capitalistic social order. Such a defense rests on prepolitically and always and everywhere "defining justice as a function of wealth maximization." This is "the idea that nearly everything works best on market logic, that economic models of behavior capture most of what matters, and that political, civic, and moral distinctions mostly amount to obscurantism and special pleading."[24] What these developments share in common is the view that the social goods upheld by the law are not changeable or malleable in the context of historical circumstance but are the products of freely arrived at rational moral choices which express natural, ultimately cosmological, forces. These two developments subsume the juridical function to the role of night watchman, the guard seated atop the enactment moralizing panopticon.

Understood in these terms, decisions by successive Supreme Court justices, eager to interpret the US Constitution through the lens of doctrinaire law and economics theory have made it increasingly difficult for advocates of the externalization thesis to justify claims for the redistribution of social or environmental goods as self-sufficient principles. Seen in historical light, contemporary jurisprudence of the kind espoused by Richard Posner gives expression to longstanding New Right efforts to undermine the idea that groups of private property–owning persons—granted a permit to act collectively, a "corporate license to operate" by the state—are potentially as or more harmful to the generalizable interests of all citizens than would be political authority. Notably on "questions of money-in-politics . . . neoliberal jurisprudence assumes a perfect (political) market and resists government intervention aimed at correcting power imbalances and anti-competitive behavior" on grounds that these distort the natural perfection that ideally unfettered economic markets express.[25]

In Sonia Amadae's view, the ideal of justice as wealth maximization central to law and economics holds that "resources are socially best allocated when held by the highest bidder, who by definition puts them to their most efficient use, has no external test and consistently determines that individuals who have more financial wherewithal are inherently more justified in owning property."[26] At the core of the new jurisprudence therefore lies a clearly holistic conception of society combined with an enactment moralizing conception of authority. Such a conception panoptically assesses matters brought before the law in cost-benefit terms, as additions to or subtractions from some ledger of the greatest good for the greatest number. Ironically, however, the new jurisprudence requires those making judgments to exercise Hayekian humility. While explicitly restricting the remit of the law to a transparent, quantifiable, economizing maxim, such jurisprudence implicitly relies upon knowledge of nothing less than the ledger's organizing principle. Law is thus the vehicle through which authority ensures that society expresses, rather than artificially contradicts, natural forces. Of course, it is implausible to regard legal pluralism or law and economics as closet avatars of benign *ecological* holism.[27] However, after having reinforced the doctrine of corporate legal personhood in the *Citizens United* case in 2010, the Roberts Court in the 2014 *Hobby Lobby* case seems indeed to recognize that the business of business is not exclusively to turn a profit but to maintain the natural order of things. In fact, the decision makes this point clearly:

> While it is certainly true that a central objective of for-profit corporations is to make money, modern corporate law does not require for-profit corporations to pursue profit at the expense of everything else, and many do not do so. For-profit corporations, with ownership approval, support a wide variety of charitable causes, and it is not at all uncommon for such corporations to further humanitarian and other altruistic objectives. Many examples come readily to mind. So long as its owners agree, a for-profit corporation may take costly pollution-control and energy-conservation measures that go

> beyond what the law requires. A for-profit corporation that operates facilities in other countries may exceed the requirements of local law regarding working conditions and benefits. If for-profit corporations may pursue such worthy objectives, there is no apparent reason why they may not further religious objectives as well.[28]

Corporations, their owners and managers, are conceived here as actors possessing a natural ethical sense not only for assessing self-interest but also for ratiocinating over right and wrong, be it religious or not. Corporations "themselves," as well as or just as do their owners and managers, confront moral dilemmas. And, implicitly, when liberated from the egregious impacts of politics acting through the law, the law promotes such rational choices. The law serves to maximize the sum total social good as it can be quantified in terms of economic units of utility. Importantly, this social good is not conceived here in humanitarian or altruistic terms. Rather, it is conceived somewhat nihilistically as the product of for-profit corporations' freedom to egoistically choose such terms or not. This decision builds directly on the *Citizens United* finding that the political viewpoints of corporations themselves are valuable as those of any legal persons inhabiting the wider public sphere. This is because "on certain topics corporations may possess valuable expertise, leaving them the best equipped to point out errors or fallacies in speech of all sorts, including the speech of candidates and elected officials." It is not for the judiciary, legislature, or executive to circumscribe any legal person's ethical sense, nor to distinguish between individual and corporate interests. Instead, the task of law is to facilitate free choices amongst all legal persons, such that it is "not uncommon" for some actors to "pursue more worthy objectives than others." It is the job of the judiciary to uphold the empirical fact of moral autonomy amongst all legal persons, with the consequences ultimately left to the play of natural forces because moral autonomy is the human expression of such forces. As Brown recognizes, so-called "neoliberal jurisprudence . . . recasts formerly noneconomic spheres *as* markets at the level of principles, norms and subjects."[29]

It is in this sense that contemporary right jurisprudence goes a long way toward fulfilling broader new right aspirations. For example, the section of the Powell memo addressing the "responsibility of business executives" asserts that these extend beyond "maintaining a satisfactory growth of profits, with due regard to the corporation's public and social responsibilities. If our system is to survive, top management must be equally concerned with protecting and preserving the system itself."[30] Of course, Powell could not have foreseen the political and ideological successes of the movement that his memorandum encouraged. However, Powell did author the *First National Bank of Boston v. Bellotti* decision that made individual and corporate speech equivalent on grounds that the First Amendment constituted a market for speech, one potentially harmed because unjustifiably distorted by the imposition of artificial restrictions from outside by political authority. Nonetheless, the impact upon politics has been comparable to that which he envisioned for the morally upstanding top management executive. The encoding of social relations as specific instances of market economic relations, where the latter are conceived as naturalistic expressions of what is ultimately a cosmologically ordained moral order, today renders illegitimate the nonmarket capacity of political authority. Contemporary jurisprudence circumvents nonmarket principles as unnatural and political authority as immoral. In this way, contemporary jurisprudence complements both electoral left neoliberal and popular neocommunitarian efforts to silence political protest against regulatory arrangements that refer to nonmarket principles.[31]

For Shamir, whose critical interpretation is worth engaging at length, the wider implications of neoliberal jurisprudence signal a historical shift in the ways that authority justifies the exercise of power. Today increasingly,

> governments relinquish some of their privileged authoritative positions and are reconfigured as one source of authority among many, in fact re-conceptualized [legally] as if they operate within a "market of authorities," placing governments on a par with private sources of authority and changing the role of governments from regulators to "facilitators."[32]

Granted coherence and support by legal pluralism and law and economics, the governance turn to which responsible corporations voluntarily contribute thus effectively marketizes the

> means of authority such as laws, rules, and regulations . . . alongside tools such as guidelines, principles, codes of conduct, and standards. . . . All are conceived [by the law]

> as instruments that are produced, distributed, exchanged, negotiated, and ultimately consumed by the host of state and interstate agencies, commercial enterprises, and nonprofit organizations that comprise the "market of authorities."[33]

This "marketizing of authority go[es] beyond specific policies to constitute a market-like representation of the very notion of authority," and frames

> moral issues through the foundational epistemology that dissolves the [artificial] distinction between market and society and, furthermore, encodes the "social" as a specific instance of the "economy." Moral considerations thus "lose," so to speak, their transcendental attributes or at least their character as liabilities and re-emerge as business opportunities. . . . The result is a shift from deontological [externally derived and politicized] ethics to [holistic] ethics that subordinate socio-moral sensibilities to the calculus of possible outcomes, to the tests of cost-benefit analyses and to the criteria of reputational-risk management.[34]

In place of the regulatory state has emerged a true catallaxy, a system of governance in which, through the law, moral authority instantiates the natural ordering principles of the cosmos. In place of deontological artificial regulation, the marketizing of authority establishes an ontologically holistic cosmological continuum. Political authority is withdrawn and displaced by genuinely moral authority. Such neo-Enlightenment, postpolitical moral authority originates within, and not from any source external to, market forces. However, this withdrawal of an artificial dualist authority and its replacement by an ontologically holist conception of authority does not represent the abdication of power in favor of some kind of unmanaged state of nature. Rather, it represents the recalibration of power in the image of the morally worthy and the strong. Such forces, channeled by the truly worthy, express the cosmos's propensity to spontaneously self-order in ways that favor the most rational of moral choosers, those who enact the survival of the fittest. Authority is no longer willing to facilitate any deontological body in response to artificial demands. Rather, authority is self-understood as the facilitator of an unfolding natural order, which is simultaneously the true moral order requiring enforcement. In this way, actors choosing, and capable of choosing, to operate in ways that demonstrably express natural forces gain and rightfully hold the most power.

Shamans of the Anthropocene?

It is in this sense that what Crouch, Streeck, Mouffe, Brown, Fraser, and Runicman describe as a postpolitical order entails the revival of moral over political authority. While such a postpolitical order is ontologically speaking holistic, it is not homogeneous. Ideologists of the electoral left and right emerge within the postpolitical order as competing vanguard parties for society's anticipated harmonization within the cosmological order.[35] The postpolitical order effectively sustains two competing iterations of moral authority. Moral authority might be justified in left neoliberal terms, as a commitment to benignly steering society's participation in the ecosphere onto a sustainable footing through policies that enhance ethical rational moral choices and so promote green economic growth. Or moral authority might be justified in popular neocommunitarian terms, as a commitment to returning the nation to greatness through policies that unleash the enterprising spirit of Darwinian natural selection and so privilege deserving winners. In either case, the generalizable interest is no longer conceived as a matter of political contest amongst social classes privileged by their capacity to exercise power. Instead, contestation over defining the generalizable interest is terminated in favor of moralistic debate amongst more or less worthy vanguards, who recognize and act on the truth that there is no outside to the system, no alternative. In place of politics has emerged a new pretense for authority, that of ensuring harmony with natural forces, be they conceived as benign and ecologically sustainably manageable or destructive and deserving of Darwinian unleashing. Left neoliberals and first and second generation popular neocommunitarians, responsible corporations and legal practitioners of the new jurisprudence, each play a mimetic role as shamans of the Anthropocene.

Political and cultural historians Thomas Frank and Steve Fraser seem to emphasize something similar when discussing the political polarization that has become the norm since, and in large part as a consequence of, the '60s. For Frank and Fraser, politics has become a theater-like arena in which a striving class of elite left neoliberal do-gooders piously dedicate themselves to a chimerical sustainable capitalism vie for power against a reactionary right wing of patriarchal and authoritarian family capitalists dedicated to winning Darwinian struggle unmitigated by reason or principle.[36] What both factions share in common is an interest in dissolving political authority grounded by differences of interest and upheld by parties contracted to dispute the spoils of social reproduction, sustainable or not. Whether Frank's and Fraser's cosmopolitan strivers or troglodyte patriarchs define the greatest good for the greatest number remains a moot point. Both seek to shape individuals in the image of true moral authority. In this light, it does seem justified to assert with Radkau that we might "speak of an ecological age . . . eco-age . . . a New Enlightenment" that first took shape in the late 1960s and early 1970s and was entrenched by the late 1990s.[37]

While it was once thought that a radical program of dismantling ontological dualism and replacing it with holism and undermining collectives organized to master nature in favor of moral voluntarism would lead to change for the better, one that would halt environmental and human exploitation, this has not been the case. Radical green-left modernist efforts to mobilize action around the externalization thesis have become ensnared in what activist-author Naomi Klein calls "a gooey corporate hug" that admits only of "positive contributions" and brooks only free-market solutions to the problems of externalization.[38] Not only market agency but the law as authority's institutionalized reflection have taken on the qualities of what Radkau calls neo-Enlightenment values. Authority no longer operates on artificial but on the holistic principles immanent in the cosmological order. These are upheld by the most ethical or courageous and so morally worthy legal persons.

Once, the defining preoccupation of political authority, artificially separated from moral authority, was the measurement, predictability, order, and the control of objective nature, albeit broadly defined to include human nature. People on the modern Left sought collectively to influence or win political authority in order to universalize conditions for freedom as autonomy. Moral authority was something understood not merely by people on the left but also on the patrician right as a form of power exerted separately from and directed at political authority. Since the late 1990s, however, the central preoccupation both of those seeking and of those in power has shifted away from petitioning rationalized authority. Authority has come to incorporate concerns hitherto, and of course problematically, regarded as the preserve of moral authority. Concerns previously relegated to the realm of moral choices incompatible with rationalized authority emerge as postpolitical demonstrations of worthiness to hold authority, demonstrations expressive of the true natural order of things. The Left demonstrates its worthiness by acting ethically to render society's participation in the ecosphere sustainable, and the right upholds an ethos that champions the survivors in Darwinian struggle. The once radical modernist ethos central to the Green Left's holistic critique of Enlightenment dualism has been transmuted into a new reenchanting myth. Again, following Radkau, with

> the demise of the great ideologies [holistic] ecology is left as the only intellectual force giving content to the new global horizon and responding to the new challenges. The very fact that . . . environmental movements repeatedly dissolve into the mainstream bears testimony to the epochal character of environmentalism: it defines the age more powerfully than even many environmentalists would like. The chameleon-like character of [holistic] ecology is proof of its vitality—as a philosophy of life and source of political legitimacy, as science and as watchword of protest [and of reaction].[41]

Holistic ecology and moral voluntarism underpinning ethical eco-awareness today offer not an essential grounding for critique but a "chameleon-like" myth of reenchantment: the individual exercise of moral choice will ensure the whole's ethical conformity with the laws of nature. If it is to be renewed as a tool that can supply meaning to people on the left, critique might better turn away from a defunct Enlightenment myth of progress through instrumental reason. The central axioms of the neo-Enlightenment age of ecology—an ontology grounded by holistic ecology and an episteme constructed around the ethos of moral voluntarism—might today be better treated in the same dialectical terms used by Horkheimer and Adorno when seeking to understand the pathologies of Enlightenment reason. Indeed, such a myth lost whatever power it had some time ago, as critical scholars such as Blühdorn and Luke, Frank and Fraser, as well as Beck and others such as Crouch, Streeck, Mouffe, Brown, Fraser, and Runciman, show. In the context of postpolitical moral authority, struggles against domination continue and these, I argue in the next chapter, might be understood in terms of a negative nonidentitarian *critique of* the holistic critique of Enlightenment dualism.

Chapter 6

Meaning Lost, Meaning Refound . . .

An Inebriate Tendency toward the Absolute . . .

Beginning in the mid-1990s and extending to the Occupy "moment" in 2011 a diverse array of new New Left groups helped to constitute a global justice movement opposed to the triumphalist market globalism that had underpinned the Washington Consensus since the collapse of the Soviet Bloc.[1] These included groups such anarcho-primitivist and neo-Luddite eco-collectives, militant urban squatters such as the black bloc, free party groups such as Reclaim the Streets, self-help groups such as Food Not Bombs, and various communal and social economy groups. Well-versed in the physical and social sciences and humanities, scholars and activists associated with such groups drew inspiration directly from the deep and social ecology and the communitarianism of the 1970s, as well as from older traditions of anarchism and Marxism. Advancing the work of the radical scholar-activists and intellectuals of the 1970s, these revivalists sought to link holistic ecological with identitarian, labor, and civil rights demands. Two important contributions, both fusing prefigurative politics with direct action protest, were ecological neo-anarchism and a simultaneously post-Marxist and postcapitalist radical communitarian localism. Just as had the original green-left modernist synthesis that took shape in the wake of what Radkau calls the first great eco-boom of the 1970s, the green-left modernist revival that coalesced in the wake of the second great eco-boom of the 1990s, wittingly or not, gave expression to deeper philosophical currents. Arguably the most influential stream within these broader currents was one that had first taken shape in French universities in the 1960s and was later exported to universities and colleges in the United States as postmodern critical theory or post-structuralism.

As Habermas had noted more than a decade prior, insofar as post-structuralism grew out of and responded to both social scientific structuralism and philosophical existentialism, its advocates often drew heavily on Heideggerian interpretations of Nietzsche's philosophy. Thinkers such as Foucault, Jacques Lacan, Jean Baudrillard, Jacques Derrida, Gilles Deleuze, Felix Guattari, and Ètienne Balibar are perhaps the most famous avatars of such post-structuralism. However, the impacts upon this new New Left were most directly exerted through the work of their epigone, Michael Hardt and Antonio Negri. Hardt and Negri's best-selling titles *Empire*, *Multitude*, and *Commonwealth* sought explicitly to unify Nietzschean and Heideggerian with older Spinozist themes. Hardt and Negri sought to combine Heidegger's interpretation of Nietzsche, through which the former founds freedom on the courage to know necessity, and in the face of that necessity to identify when and with whom it is appropriate to celebrate the oneness of being as presence, with a Spinozist understanding of freedom as the product of seeking joy in acquiescence to the same necessity. Indeed, at the height of the revival, and echoing Habermas's critique of post-structuralism as an aporia for critique, neo-Marxist Tom Nairn derided the new wave of scholar-activists for partaking of the Spinozist "philosophical elixir of oneness" and so, wedding an "inebriate tendency towards the absolute [with] a necessarily total, secular faith fusing conceptual satisfaction and moral-political guidance." Another individual with firsthand experience of the '60s, Barbara Epstein, also raised such suspicions in her detailed study of prefigurative direct action movements from the 1970s into the early 1990s. For Epstein, prefigurative nonviolent direct action protest brought into engagement with the postmodern political culture of the globalization era longstanding American and so longer standing continental Lutheran Protestant beliefs in "conscience-based subjective resistance to authority."[2]

Although some neo-anarchists, especially David Graeber and John Zerzan, strongly reject suggestions that their ideas are in any way related to post-structuralism or are expressions of postmodernism, radical communitarians such as the writing partnership known as J. K. Gibson-Graham openly avow such influences.[3] The neo-anarchist leaning scholar-activists of the revival generally concur with Graeber, insofar as they advance ideas associated with nineteenth- and twentieth-century philosophical anarchists (including, incidentally, Henry David Thoreau, a key source for the deep ecologists and Murray Bookchin, the main proponent of social ecology).[4] Such neo-anarchists seek to mobilize action informed by the "basic principles of anarchism—self-organization, voluntary association, mutual aid," which are taken to be normatively important because they constitute "forms of human behavior . . . assumed to have been around about as long as humanity."[5] These principles are opposed to implicitly newer and less desirable forms of behavior characteristic of modernity. Alternately, radical communitarians tend to concur with J. K. Gibson-Graham and take post-Marxist, postcapitalist inspiration directly from post-structuralism. In the radical communitarian view, capitalist modernity, and its alternate productivist Marxism, seem "quite distant from both [the] *personal and social* transformation" that emancipation from domination requires. Indeed, "Marxism has produced a discourse of Capitalism that ostensibly delineates an object of transformative class politics but that operates more powerfully to discourage and marginalize projects of *class* transformation."[6]

Both the neo-anarchists and radical communitarians derive their activism from a narrative of declension. Just as had the deep and social ecologists and communitarian revivalists of the 1970s, advocates of this narrative regard not only the instrumental rationalism that defines modern artificial political authority but also the modern Left that opposes it as representatives of a fall from some socio-environmentally benign, harmonious communal social form. In this view, prefigurative actions and diverse community economies provide a normative guide that serves to "foreground the way that assemblages of objects and practices . . . both enact the economy and contain the performative potential to constitute the 'economy' as other than itself," that is, as other than the neoliberal global competition state.[7] However, while neo-anarchists urge action based on the "rejection of the state and of all forms of structural violence, inequality or domination," radical communitarians do not reject the state, so much as they regard it as "an ever-present danger that calls forth vigilant exercises of self-scrutiny and self-cultivation—ethical practices, one might say, of 'not being co-opted.'"[8] Neo-anarchists also see danger in the state's rationalized authority, and tend to follow Peter Kropotkin's nineteenth-century advice to operationalize political action on the basis of

> a principle or theory of life and conduct under which society is conceived without government—harmony in such a society being obtained, not by submission to law, or by obedience to any authority, but by free agreements concluded between the various groups, territorial and professional, freely constituted for the sake of production and consumption, as also for the satisfaction of the infinite variety of needs and aspirations of a civilized being.[9]

As discussed above, the moral volunteer is justified in this role because neo-anarchists believe such a monad to have been around about as long as humanity, and so must embody universal desires for civilized being. The agent for society without government is the autonomous moral volunteer. In the radical communitarian view, a similar moral volunteer chooses to act on the ethical obligation to build and sustain a planetary "ecology of voluntary associations."[10] For radical communitarians, the agent is the ethical actor who mobilizes a

> "politics of the subject" . . . a process of producing something beyond discursively enabled shifts in identity, something that takes into account the sensational and gravitational experience of embodiment, [a moral politics] that recognizes the motor and neural interface between self and world as the site of becoming of both. If to change ourselves is to change our worlds, and the relation is reciprocal, then the project of history making is never a distant one but always right here, on the borders of our sensing, thinking, feeling, moving bodies.[11]

The neo-anarchist's principled rejection of the state and the radical communitarian's politics of the subject seem to inform a hyper normative conception of an autonomous citizenry that rejects outright the transferal of political authority to the state in exchange for some quantity of public welfare. In place of artificialist justifications for the state as the least-worst option for organizing authority in relatively complex, populous, and geographically expansive territories that must persevere in an anarchic international system, these revivalists seek to establish a deliberative community that self-formulates and applies the rules needed to address social and environmental problems as if the world were truly a global village.

The shared trope of both neo-anarchist and radical communitarian thought is in this sense the decades-old green-left modernist idea of prefigurative politics. Both neo-anarchists and radical communitarians agree with the deep and social ecologists and the communitarians of the 1970s that actors who choose to act ethically provide a prefigurative alternative example to the rest of society of how authority can work without politics. Following from this, the free and autonomous communities that radical ethical actors constitute serve both to delegitimize and replace the globalizing competition state and its equally oppressive and outdated opposition, the modern Left. The belief is that effective political agency requires "an 'infinitely demanding' ethics that entails carving out authentically political space at an 'interstitial distance' from power . . . [aimed at] 'exposing, delegitimizing and dismantling mechanisms of rule while winning ever-larger spaces of autonomy from it.' "[12] Or it requires "a new political project of configuring . . . a space for thinking about 'noncapitalism' " and the development of a non-capitalocentric "discourse of economic difference."[13] Prefiguration in this guise expresses a "performative ontological politics" that operationalizes an "explicitly political interest in imagining and enacting other worlds."[14]

This said, neo-anarchists and radical communitarians do diverge somewhat on how they interpret the logic of prefiguration. For the former, the well-known anarchist desire to eschew theory in favor of practice or direct action takes precedence. For example, neo-anarchists commit to prefiguring the ends of political action on the assumption that ethical actions guided by moral choices reflect or are "somehow equivalent, to the ends."[15] Neo-anarchists aim to create new political conditions that are intended to replace existing political conditions by literally prefiguring them, such that "an action is prefigurative when it fulfils certain conditions in the way in which it is performed." Examples include conditions that arise when establishing a protest camp on radical egalitarian lines, or organizing the distribution of otherwise wasted food.[16] For Graeber, neo-anarchism is

> a movement about reinventing democracy. It is not opposed to organization. It is about creating new forms of organization. It is not lacking in ideology. Those new forms of organization *are* its ideology. It is about creating and enacting horizontal networks instead of top-down structures like states, parties or corporations; networks based on principles of decentralized,

> non-hierarchical consensus democracy. Ultimately, it aspires to be much more than that, because ultimately it aspires to reinvent daily life as whole. But unlike many other forms of radicalism, it has first organized itself in the political sphere—mainly because this was a territory that the powers that be (who have shifted all their heavy artillery into the economic) have largely abandoned.[17]

In Graeber's view, prefigurative practice encompasses not only setting an example of how things might be done in another possible world but also simultaneously living the example of that possible world. In this sense at least, neo-anarchism echoes Heideggerian concerns to privilege involvement, or doing being in the world (*Bewandtnis*), over artificially challenging the world (*Herausforden*) through, for example, technology or constitutional rules.

For the radical neocommunitarians, theoretical depth is paramount. Although reliant on post-structuralist philosophical tropes, the radical communitarian idea of prefiguration hews more closely to older Spinozist concerns with channeling the holistic oneness of the universe through moral voluntarism. Radical communitarians privilege a theoretically grounded "performative ontological politics" that situates ethical practices

> alongside so-called capitalist activity as being on a single plane in which value, importance, and driving dynamism need to be empirically investigated and creatively theorized, rather than read off from an economics textbook [that, as a modern project can offer only] structurally determined or logically derived dynamics of economic interaction.[18]

The idea is to "break with any commitment to essentialist ontological structures that participate in generating social phenomena [because] to instate diversity and overdetermination as ontological givens is to succumb to the beguiling temptation of essentialism."[19] The mere act of undertaking performative ontological politics reveals to those who choose to embody such a politics, and to the wider world, how such ethical activities—green consumption or building a workers' or producers' cooperative, for example—actually reconfigure the "neoliberal economy" as other than itself.[20] The radical communitarians evoke the majestic plural as an aid to realizing the fantasia of oneness: "As we begin to conceptualize contingent relationships where invariant logics once reigned, the economy loses its character as an asocial body in lawful motion and instead becomes a space of recognition and negotiation."[21]

The Hangover . . .

Following over a decade of the green-left modernist revival, the ongoing global financial crisis, and in the wake of the Occupy protests' collapse in late 2011, a number of other scholars and activists once associated with it began to rethink such holistic and voluntarist assumptions. Especially focusing on the commitment to prefigurative politics, Blair Taylor argued in 2013 that much neo-anarchist thought is both ill-informed and misguided. Setting the revival in the historical context of the birth of electoral left neoliberalism and the governance turn, Taylor convincingly argues that, in contrast with the aims of its proponents, neo-anarchist prefigurative politics is, in practical application, "heavily focused on personal consumption," and offered "a micropolitical politics deeply wary of power, a moralistic and often personalistic economic analysis that emphasized both consumption and alternatives, and a predilection for tactics in lieu of political strategy."[22] It is worth quoting Taylor's analysis at length:

> This political subjectivity discursively echoes two important tropes of contemporary capitalism: first, endless rebellion against the status quo, either in the creative destruction of neoliberalism or the logic of modern advertising and fashion; and second, a notion of autonomy which presumes such a distance from power is possible. Thus, it is not surprising to find other affinities between neoanarchism and [left] neoliberalism. Both share a drive to dissolve state power, in one through the creation of autonomous spaces, in the other via privatization and markets. The commitment to consensus and hostility to ideology is consonant with the post-political tenor of [New Right] "End of History" arguments . . . the logic of "exodus" mimics the fait accompli of capital flight; voluntary social services like Food Not Bombs fill the vacuum of state entitlements with charity; activist projects like the Independent Media Centers rely on a DIY ethos of voluntarism that shadows the disappearance of professional careers in fields

> like journalism into the democratic yet unpaid "blogosphere"; "infinite responsibility" mirrors the neoliberal discourse of personal responsibility; the concomitant micropolitics of voluntary simplicity and ethical consumption offer a constellation of guilt, asceticism and expensive "ethical" commodities which render ecological practice a profitable form of left austerity.[23]

Picking up the thread of debates undertaken between neo-Marxist Slavoj Žižek and neo-anarchist Simon Critchley at the height of the Occupy moment, Taylor identifies a host of similarly close "affinities between neoanarchism and neoliberalism."[24] For Taylor, "despite the rhetoric," the neo-anarchist emphasis on "endless personal rebellion [and] autonomous spaces" actually serves the neoliberal emphasis on "dissolving state power" and the "privatization and markets," such that prefigurative "boycotts, alternative products and consumer activism not only does not threaten capitalism, but in fact reinforces it by creating new markets." Meanwhile, neo-anarchists continue to define themselves "in opposition to a form of state power that capital has largely already defunded if not defanged.[25] Ultimately, in Taylor's analysis, neo-anarchism occludes rather than undermines structural biases within market economies against concerns like labor rights or ecology. This is because as an ideology, neo-anarchism closely resembles what I have called the corporate social and environmental responsibility movement, which in fact

> speaks the language of social movements, addressing ethical concerns while simultaneously insulating itself from critique. . . . [Neo-anarchists' influence began to] decline at the same time that their signature issues—ecology, labor, human rights—were commodified, incorporated into a newly emerging discourse of ethical capitalism wherein consumers can vote with their dollars in a seemingly democratic and noncoercive market that claims to give people what they want, without the messy business of politics.[26]

In short, the concept of prefiguration central to the green-left modernist revival echoes both left neoliberal and popular neocommunitarian emphases on an all-encompassing holism, on consumer choice over regulation, on corporate owners' and managers' ethical, as opposed to their political, responsibilities, and on community deliberation rather than politically enforceable internalization of externalities. The green-left modernist revival thus seems premised on the same ontological commitments and ethical priorities as are left neoliberalism and popular neocommunitarianism.

In this light, it seems fair to say that the green-left modernist revival that began in the late 1990s offers few if any grounds for a renewal of meaning on the left in the early twenty-first century. Neo-anarchists such as Graeber and Zerzan and radical communitarians such as J. K. Gibson-Graham unreflexively articulate seemingly Adornian *modernist* ("postmodern" in Epstein's terms) sensibilities, yet do so within a post-political context that depends on such sensibilities themselves. Until the mid-1970s, modernism had been subsumed to what critical theory identified as *modern* political culture. It was, after all, modern political culture—the "authoritarian personality"—that sustained the dualistic, reenchanting Enlightenment myth of progress through rational mastery. Modern political culture simply did not value, recognize, or respect modernism. However, by the early twenty-first century, such modernism had become the driving force within the engine of modernity. Left neoliberal and popular neocommunitarian right parties both appealed to risk-aware yet precariously situated individuals through the moralized politics of fear. As Beck's subpolitics in risk society thesis and the postpolitical literature shows, the modern myth of progress through rational mastery has been rendered all but meaningless. The rational egoistic utilitarian calculator in whose image the powerful sought to mold individuals has been transformed. No longer premised on dualistic control over human and nonhuman nature, the twenty-first-century utilitarian calculator is an autonomous self-realizing social atom, seamlessly self-integrated within an all-encompassing system. What the radical green-left modernist revivalists failed to notice in the lead up to Occupy was that their social ecological and communitarian forebears in the 1960s and 1970s had offered holism and voluntarism as an external—nonidentitarian—critique of the system and the myth of rational mastery that sustained it. Unreflexively adopting their holistic critique of Enlightenment dualism, the revivalists' advocacy of holistic ecology and moral voluntarism served decades later to reinforce rather than critique the holistic neo-Enlightenment myth of reenchantment, be it cast in benign terms of an ecologically harmonious and sustainable lifeworld community or in malignant terms of the greatness achieved by winners through competitive struggle. The revivalists seem to mistake a historical contingency—a system of authority premised on the rationalized authority of the state—for a necessity—the use by the powerful of whatever rationalizing premise is at hand to justify authority and so perpetuate the given order.

Critique for another Time Past

In chapter 1, I engaged Geuss's essay "The Loss of Meaning on the Left" to highlight some of the critical theoretical ramifications that Adorno drew from Marx's nineteenth-century inversion of Hegelian philosophy. Hegel's insight was that ideas are the products of history, and that awareness of this fact provides the engine of modern progress toward universal freedom. Marx inverted the Hegelian idealist schema into a structural materialist argument that ideas and the actors that create them, indeed the entire real world, are shaped by the historical, political, economic, and cultural contexts in which they are constituted. For Marx, *collective* experience and recognition of this fact is the engine of modernity. Hence, his or at least Engels's prediction that once the proletariat gained insight into the workings of the state-market nexus central to capitalism, they would rise up to overthrow such a system and replace it with a system of rational mastery over nature, communism. Adorno's contribution to Marx's Left Hegelianism was to elaborate his recognition that the collective effort to exert rational mastery over nature with the aim of expanding human freedom was bound to generate grave pathologies. Also examined in chapter 1 was Adorno's effort to develop this insight through recourse to Nietzschean philosophy. Nietzsche's central insight into modernity was different from and opposed to that developed by Marx. For Nietzsche, through exceptional exercise of the will, *individuals* can and should exempt themselves from the engine's rhythmic pulse. An individual who chooses to embrace a more-than-modern value system must reject the constraints of collectively justified rationality, the rules that govern the engine of modernity. In Adorno's thought, the modernist individual instantiates Nietzsche's ethical vision, that of making one's home within modernity. For Adorno, the modernist exploits the cracks and fissures, the weaknesses and contradictions that the rationalizing system leaves open or creates. For Adorno, these cracks and fissures make possible the fleeting liberations of transgressive art and aesthetic experimentation. To make a point often neglected by his more outré interpreters, such modernist liberation depends upon public education for citizenship, and those elements of the welfare state that facilitate and subsidize what might today be called quality of life and self-improvement-oriented "free time."[27]

However, it was Nietzsche's methodological individualism that followers of Heidegger and his own post-structuralist followers exported to the United States in the 1990s. As we have seen, such ideas were avowedly or unwittingly incorporated into the radical green-left modernist revival. As Adorno recognized, however, Nietzsche's modernist sensibility is simply not amenable to the formation of a collective radical political project. This is because such a sensibility encourages its adherents to interpret modern political authority in a specific way. Marx's structural materialism is inverted and reduced to the expression of a banal truism: when actors act on ideas they—ontologically and epistemologically, materially and ethically—constitute the historical, social, and genuinely ecospheric contexts in which they must live and, in doing so, change those contexts and themselves. Linking Nietzsche back through Heidegger to Spinoza, an ostensibly ethical actor confronts potentially world-changing choices. By exercising the will and making *the* moral choice (ethically becoming "other than"), such an actor does not merely alter the ontological and epistemological, the material and ethical realities that govern the engine of modernity but the real future unfolding of them. Individual being "becomes" the unfolding of history as the morally worthy build communities of the righteous. The majesty of Spinozist universal oneness and Heideggerian foundation-making are lived out in everyday life, regardless of scale, location, or external conditions. The will of the ethical actor alters both the quality and trajectory that the engine of modernity would have otherwise traced out. Yet, in the absence of structural considerations such as scale, location, and external conditions, the result is more than likely what Nairn derides as a "rapturous merging . . . the Spinozan framework identifies everything with everything else, including politics, thus restor[ing] a universalist vocation to the intelligentsia."[28]

In these terms, the Spinozist Heideggerian influence on green-left modernism suggests a worldview that closely resembles that embraced by left neoliberals and popular neocommunitarians, even if now conceived structurally, from the bottom up, rather than in enactment terms, panoptically. This is a worldview built on the naturalistic fallacy, which, to reiterate, presumes a desired social state can be comprehensively explained reductively in terms of unobserved natural properties or forces. In the case at hand, ethical actors make the moral choices necessary to build other possible worlds. These worlds prefigure in the here and now a benign ecologically and socially harmonious society. Such other possible worlds are assured simply because it is believed that such prefigurative communities express a whole that would be pristine but for artificial, pathological, impositions. For radical communitarians, prefigurative politics begins from the plausible but highly improbable belief that as actors engage in ethical activities, "seemingly fixed and stable essences of social identity are reframed as more or less successful processes of becoming," with the result that unethical "capitalist dominance"—albeit at an unspecified scale and in relation to an unspecified political issue—can no longer exert unethical "performative effect[s]."[29] Prefiguration of this order allegedly "reconfigures . . . capitalocentrism," because actors' "*understanding of power* enlarges the field of their own effectivity" such that moral choices to act ethically serve as "a site of becoming, and as the ground of a global politics of local transformations."[30] Alternately, Graeber's claim that prefigurative decentralized forms of acting and doing are actually the movement's ideology, and not demonstrations or representations of it, also seems suspect. This claim too is premised on the naturalistic fallacy. Graeber's prefiguration draws on a similarly intuitively plausible but practically improbable belief that the kinds of emancipatory political structures sought by Green-leftists—at any scale, over any local, regional, or global area, or of any desired consequence—can and do evolve organically. For him, the political ends are "immanent in the anti-authoritarian principles that underlie practice" such that

> a rich and growing panoply of organizational instruments—spokescouncils, affinity groups, facilitation tools, break-outs, fishbowls, blocking concerns, vibe-watchers and so on—all aimed at creating forms of democratic process that allow initiatives to rise from below and attain maximum effective solidarity, without stifling dissenting voices, creating leadership positions or compelling anyone to do anything which they have not freely agreed to do.[31]

In this view, society is conceived as an uninterrupted moral continuum that extends from activists' self-conscious embedding of the ethical self in prefigurative practices, through the polity as a whole, to the constitution of what is taken to be in principle an anti-authoritarian cosmological order. The dialectic between *immanence*—the radiating outwards of a normatively desired ethos—and *transcendence*—the incorporation of such an ethos into that of the polity, such that society comes to operate on new principles—is understood as a process of being as becoming. It is of course true in a materialist ontological sense that any prefigurative practice does indeed alter the material constitution of the world into the future. But, unless some appreciation of historical and political context, scale, and quality can be attached to such prefiguration, this insight is, politically speaking, banal. Moreover, the impulse elides Nietzsche's own express rejection of the legitimacy of any collective lessons to be drawn from the modernist ethos going forward, political or otherwise. Nietzsche's is a philosophical call to individuals to be better in the aristocratic sense that he defines it. In this respect, both neo-anarchists and radical communitarians have quite a lot to say about actors' ethical motives and moral choices to become other than: other than authoritarians, other than the state, other than mere voters, other than neoliberals, other than capitalists, other than party apparatchiks, other than vehicles for capitalocentrism, and so on. Indeed, it has become a staple of detractors that prefigurative politics is other than a lot, but not actually *for* anything.[32]

Influential streams of environmental political theory have responded to some of these problems inherent in the green-left modernist revival. Under the banner of sustainable materialism and the politics of everyday life, these approaches seek to build pragmatically on the Heideggerian Spinozist themes elaborated by new vitalist intellectuals such as Jane Bennett. Bennett situates her "vital materialism . . . in the tradition of Lucretius, Nietzsche, Whitman, and Deleuze" as well as Hans Driesch and Henri Bergson, "who share with me an ontological imaginary in which processes of becoming always vie with actual beings . . . [and to construct a] 'materialism' adequate to the virtual complexity of life."[33] The scaleless, placeless, and ahistorical truism that is ontological politics is by virtue of the act of individual willing, rendered adequate to addressing the complexities of a political order that facilitates the production and maldistribution of social and environmental harms. Drawing directly on new vitalist ideas, David Schlosberg and Romand Coles attribute their own pragmatic "sustainable materialism [to] a venerable tradition of thinking," which looks a lot like Heideggerian Spinozism. For Schlosberg and Coles, "we are . . . far more being, or material becoming, than thinking." "We" recognize that "work to transform our everyday lives, economies, polities and the values with which they are intertwined—will require political interventions at the level of bodies and material practices." This said, however, and unlike Bennett or

J. K. Gibson-Graham, Schlosberg and Coles do not go so far as to argue that prefigurative "movements provide the sole—or necessarily the best—answer to the host of political, economic and eco-logical problems faced in contemporary life . . . [nor] represent the singular or even majoritarian evolution of new environmental movements." Rather, their sustainable materialism is soberly conceived as a supplement to "electoral or policy-focused politics," wherein the "focal point is not to organize to lobby or vote for change [but] to literally embody that change, and to illustrate alternative, more resilient, and more sustainable practices and relationships."[34]

For John Meyer and colleagues, a pragmatic "politics of everyday life" that has no single theoretical source is interested in "how human experience is interwoven with material flows, nonhuman beings, ecological systems, and technological infrastructures" and in how activism oriented to altering those experiences can foster new ones. For Meyer, critics need to focus on how the interweaving of everyday experience with material flows might "resonate" with citizens. The task then is for critical activists and organizers to pick apart such resonance and demonstrate how the embedding of the self in "everyday life," what Bennett and J. K. Gibson-Graham would call "ontology," is in fact constituted in unsatisfactory or suboptimal ways in relation to green-left ends. Meyer therefore faults Bennett and other advocates of the new vitalism for focusing "disproportionally . . . upon ontology" to the exclusion of politics.[35] Everyday life practices—such as owning property, driving, or household consumption—are to be exposed as the products not of what Bookchin may have called "first nature" but of "second nature," as artifacts of human intention. Against arguments from ontology, the politics of everyday life aims to offer intellectual resources to scholars and activists that will allow them to bring second nature into alignment with the demands of sustaining first nature. That is, it aims to engage the everyday and so, it is hoped, build political momentum for change by convincing citizens that the world can be other than it is.

Avowed neo-anarchist and Occupy veteran Micah White similarly seeks to mobilize political agency from below from within an unalloyed Spinozist frame. For White, the political failures of direct action protest are too apparent to ignore. Yet, in his view, this is due to the lack of attentiveness by activists to the spiritual, indeed cosmological, dimension of politics. For him, "Occupy was a spiritual experience" and led to an epiphany of sorts:

> The goal [of the political agent] is to ascend from material to spiritual and internal to external attaining an understanding of each theory and a sense of when each is most appropriate. Most of contemporary activism is voluntarism. Voluntarism means focusing on human actions—getting people out into the streets to effect change. But there's also structuralism, the theory that activism stems from fluctuations in economic or political systems rather than individual protests; subjectivism, which depends on epiphanies via mental enlightenment and messages spread through social networks (as Occupy's influence expanded with hashtags and well-timed Tweets); and theurgism, which places responsibility on the supernatural.[36]

White wants to abandon materialism for an amorphous, agentless new spiritualism that will unite people from across the political spectrum in some kind of polymorphous metaphysical group hug. White seeks to harness the spiritual commitment of activists to a utopian dream. He seeks to spiritualize mundane tasks such as involvement in the electoral politics and administration of democracy through membership on public boards, for example, both hitherto eschewed by neo-anarchists. For White, building social movements and engaging in electoral politics is the only way forward. Yet, his enthusiasm for harnessing the creativity of people across the political spectrum subsumes real political distinctions beneath the Spinozist yearning for oneness. White and others such as Schlosberg and Coles and Meyer seem to forget that when push comes to shove, Left and Right do in the end hold fundamentally opposed worldviews—the former supporting universal freedom, the latter hierarchy.

My criticism here is that, when constituted through the lens provided by Spinoza and Heidegger, directly, indirectly, or pragmatically, radical green-left modernism is unable to provide a political rationale for evaluating the appropriateness or efficacy of any of this ethical agency and moral choosing to be other than. And a political rationale for evaluating action is precisely what is required to bring meaning back to the radical Enlightenment project. Translating the problem into Heideggerian jargon, it might be said that green-left modernism takes being in the world as an ethical force (*Dasein*) to express authentic as opposed to inauthentic being. Authenticity lay in the other than society that the green-left modernist revivalists seek to prefigure in the here and now. As has been shown, however, such authenticity is an article of moral suasion arrived at amongst ethically committed communards. Arguments might refer to the truth of scientific ecology, of religious or quasi-religious spiritual commitment, of the warmth and camaraderie found in alternative lifestyles or small groups where face-to-face interactions prevail, or that humans and indeed other primates are essentially communalistic and not competitive animals, for example.[37]

Yet, as Adorno already suggested against Heidegger's arguments from ontology, it must be remembered that efforts to use Nietzschean modernism as the basis for such a distinction make the prepolitical ethical actor the final and only arbiter of the authentic. And, as Nietzsche himself saw, this distinction helps to make raw power the ultimate value. Authenticity so understood can offer no recognizably collective justification for political will formation oriented to reconstituting the boundaries of political freedom. For Adorno, such modernism can offer only "the insight into [one's] entanglement and the happiness of the tiny freedom, which lies in the recognition as such." Moreover, as I point out in chapter 1, well-known is that Adorno expressed considerable dismay on witnessing his neo-Marxist appropriation of Nietzschean modernism being thought in political terms at all. It may be for similar reasons that early proponents of deep ecology such as Nathan Kowalski had by the 2010s come to regard it as a "spent force," while it is definitely the case that Bookchin declared social ecology "problematic" at around the same time because he recognized it to be the historically outmoded expression of radicalism for another time past.[38]

The revival of green-left modernism tends to replicate postpolitical assertions of authority, yet does so through powerless but pious activism that is informed by practical reasoning on the model of structural moralizing. As did the original green-left synthesis of the 1970s, revivalist green-left modernism settles the ontological question prior to politics, and does so in favor of a whole that is coextensive with individual actors' moral choices to be ethical. This is ethics on the philosophical model of structural moralizing, insofar as the whole is conceived as if it could be unproblematically rendered coextensive with ethical individuals' and communities' experiences in the world. The moral choices made by authentically ethical actors, by sheer act of moral willing, make the desired other world a reality in the here and now. But settling the ontological question holistically and grounding responsibility for its unity in the moral volunteer renders historical context, political structure, and institutions invisible or irrelevant. To begin from the ethical concept of an ecologically whole society that is the product of individuals' moral choices is simply to implicitly privilege a favored morality that should characterize all of society. This move reconfigures the answer to the ontological question (the social whole conceived in terms of an ethical actor's freely arrived at choice to do this and not that) as an abstract moral code (subjects should pursue what they, and implicitly we, think is good and right, in harmony with ecospheric processes).

As a contemporary iteration of longer standing philosophical tropes, green-left modernism moves from the empirically wrong but intuitively appealing structural moralizing assertion that the truly benign natural condition of freedom that moral actors are capable of realizing can and should be exhaustively embodied in what is in fact—by virtue of scale and complexity at the very least—an amoral political institutional structure. This is a prepolitical ideal of freedom and is undoubtedly a powerful and attractive existential experience in modernity, as thinkers from Hegel forward note.[39] However, as Geuss has more recently suggested, grounding demands for justice in an abstract and, I add, simultaneously naturalistic ideal of freedom as autonomy is an inherently problematic move for radicals and reformists alike. Just as did the modern Left when buying into Enlightenment's reenchanting myth of progress through rational mastery at the height of the Fordist-Keynesian welfare state's Cold War power, the green-left modernist revival that emerged in the 1990s buys into, rather than offers a critique of, the reenchanting myth of neo-Enlightenment. In a neo-Enlightenment age of ecology, this revival mobilized around an outmoded holistic critique of Enlightenment dualism and is for this reason radical critique for another time past.

Put differently, in modernity, philosophical reflection and political analysis, moral obligations and political obligations, manifest on two different and often irreconcilable planes. The green-left modernism that was first synthesized in the 1970s and revived in the late 1990s treats political contests as if they are always and everywhere like moral disagreements. But, as put by Williams, moral disagreements are discursively redeemable disputes about ethics, beliefs, norms, and values, indeed, even about ontological commitments. Winning a moral disagreement means persuading others that you are right, true, correct, or good. Political disagreements are about what should be done in relation to or through the authority that those in control of institutions can and do exert to respond to the first political question. The problem for radical green-left modernists is that political disagreements are always and everywhere of an applied quality and involve the exercise of power in context, and so through relatively large institutions and legal practices that have a distinct history. In the United States in the early twenty-first century, being right is irrelevant. Winning a political disagreement means assuming a degree of authority hitherto beyond one group's grasp, which implies prizing it from the hands of some other group using political power.[40]

The central concerns of green-left modernism are rightly with eliminating the unjust production and distribution of social and environmental harms. However, in the contemporary United States, there are not only harmers and the harmed but also a mediating structure and institutions that, more often than not facilitate the domination of some by others. This is not to suggest that political disputes are mere conflicts of interest amongst rational actors. Rather, it is to affirm that political contests are bounded conflicts over the authority to organize the institutions and rules that will prevail in a large, heavily populated, and highly complex polity. That is, political contests are conflicts over the boundaries of political freedom as a concomitant of civil peace. Desirable or not, such disputes are unavoidably today conducted against the backdrop of artificial structural and institutional constraints and facilitators. This is a point not lost on the New Right scholars and activists who since the 1970s have worked assiduously to take control of all manner of institutions, from local school boards to state government and all three branches of the federal system. Aside from obvious differences in funding, what distinguishes the New Right from the new New Left is not the desire to do away with rationalized political authority. Indeed, both seem to seek that same end. Rather, it is the centuries old difference marked out by the Right's commitment to hierarchy and particularism over the Left's dedication to egalitarianism and universalism.

From Occupy to the Trump Administration

As pointed out above, it was Epstein who in the early 1990s first elaborated a set of suspicions about prefigurative politics and direct action protest. In the context of a resurgent New Right that in the 1980s had begun "presenting itself as the genuine force for democracy," Epstein suggested that advocates of spontaneous and ephemeral "prefigurative politics and utopian democracy" tend to ignore or elide a fundamental empirical fact of politics in modernity: "It is the process of building democratic and revolutionary organizations and institutions . . . that is crucial to social transformation."[41] Hence, she found that many activists simply failed to see that prefigurative politics were an insufficient vehicle for extending the influence of green-left modernism "into the institutions of mainstream society . . . the Democratic party, trade unions, mainstream churches, and other such organizations."[42] More pointedly, Epstein argued that

> There is an assumption running through the direct action movement that constructing an egalitarian, nonviolent society requires abolishing power relations and doing away with conflict. . . . It is [wrongly] assumed that conflict means violence, that power means domination, and that all forms of hierarchy are bad. These assumptions run so deep that they are rarely if ever examined.[43]

A little later, social ecologist Bookchin perceived the unwillingness of many people on the left to countenance the possibility that a historical dialectic of critique, assimilation, and recuperation was at play as the globalization process unfolded. For him, the constant cycles of radical protest, co-option, and commercialization of radical ideas demanded that green-left movements rethink the presumptions with which they worked. They perhaps had been drawn into Adorno's false nonidentity thinking, Luke's artificial negativity. This realization led Bookchin to reject philosophical anarchism. Bookchin claimed that such ideas were the preserve of "individualists whose concepts of autonomy originate in a strong commitment to personal liberty rather than to social freedom" and were the unhelpful "radical celebration of Nietzsche's all-absorbing will."[44] Bookchin effectively denounced the modernist dimension of radical green-left thought and agency that in the 1970s he had done so much to synthesize. For him, political agency derives its power not from a metaphysics of freedom that automatically converts individuals' ontological commitments and ethical priorities into harmonious ecospheric interrelations, but from political commitments to organized planning rooted in popular participation. Bookchin and other mid-twentieth-century anarchists, such as Noam Chomsky, understood

> anarchism as an ideal rooted in Enlightenment assumptions about the rationalization of social life and individual behavior. Their respect for the organizing power of reason was related

> to [the fact that they] did not fall prey to an excessive love of spontaneity, placing a considerable amount of stress on the importance of conscious and considered social planning.[45]

While Bookchin's critique won him few supporters at the time, it was arguably his long-term commitment to Enlightenment principles—and to truthfulness in the face of domination—which allowed him to think through what he derided as the dissolution of green-leftism into celebratory Nietzschean willfulness.[46]

Alongside that developed by similarly experienced scholars and activists, both Epstein and Bookchin's analyses of nonviolent direct action networks with close ties to deep and social ecology did influence the slowly building proliferation of calls to reconsider prefigurative politics amongst neoanarchist revivalists of green-left modernism. Central to the debates provoked by such calls has been the development of a conceptual and practice-based distinction between prefigurative activism and public organizing to resist domination. The longstanding status of appeals to recognize distinctions between (prefigurative, ethics-based) activism and (strategic, collective political) organizing is important insofar as it seems to dovetail with another, possibly longer standing issue amongst anarchists: lack of historical memory. This tendency of anarchists in particular to forget the past was actually first raised in the early 1960s in the context of tensions within the student and civil rights movements, was reraised by Epstein in the 1990s and came up again in the wake of Occupy. As Graeber himself put it in 2013, there seems to be "in the US at least . . . a problem with passing on the wisdom of past generations of thinkers and activists."[47]

Similarly, for Astra Taylor, the uptick of campus and online activism in the wake of Occupy seems grounded by a historically uninformed and politically unreflective idealism. For her, activists are unaware or unwilling to learn from the experiences of those who came before, and cling to beliefs that "even when disconnected from any coherent strategy," ethical action and moral choices, to whit, prefiguration, "can magically lead to a kind of societal awakening."[48] Perhaps longtime green-left activist L. A. Kaufmann puts it most clearly when arguing that since the 1970s many have worked to redeem the "promise" of consensus-based prefiguration "again and again like dogma. But let's face it: the real-world evidence is shaky at best."[49] Or, rather more bluntly, as Alex Prichard and Owen Worth argue, "occupying can only provide an incoherent form of resistance at a macro level."[50] Arguably then, it has been a commonplace amongst anarchist and more recently neo-anarchist scholars to bewail the fact that those drawing on the tradition repeatedly fail to learn from the experiences of past anarchist movements and debates.

What seems different in the wake of the collapse of Occupy is that interest in the distinction between what is essentially ethical or moral suasion and political contest has emerged as the defining feature of a new wave of support for radical and reformist politics amongst the so-called educated precariat.[51] Such interest became clear in the ultimately failed efforts of presidential hopeful Senator Bernie Sanders during the 2016 election season, in which Sanders and others including congressional candidate Zephyr Teachout allied with groups such as the Working Families Party.[52] It also became clear in more radical efforts within social-democratic, socialist, Marxist, labor-green, feminist, and LGBTQI debates to shift the focus of public discourse away from preoccupation with morality, ethics, and agency to issues relating to structure, class, and solidarity.[53]

For Chris Dixon, any effective politicization of the prefigurative idea must involve "running organizing campaigns *in combination with* building counterinstitutions" that can work "against-and-beyond" what neo-anarchists perceive to be an illegitimately authoritarian system.[54] For him, contemporary neo-anarchists should no longer see themselves as engaged in some kind of exercise in "purity-based-politics," but in political "*experiments* rather than [life] forms that are somehow universally correct."[55] Others such as Andrew Cornell go further. In discussion with former members of Movement for a New Society, active in the 1970s and 1980s, Cornell recognizes that a key reason for that organization's self-dissolution was painful recognition by its members that prefigurative politics had failed. Prefiguration simply did not appeal to those who did not share upper-middle-class backgrounds or, more pointedly, were not white. Such prefigurative politics were fueled by an unreflexive celebration of spontaneity and haphazardness for their own sake and led to stagnation.[56] Recognizing these problems, as Cornell puts it, "meant accepting some degree of both centralization and hierarchy—words that unsettled [Movement for a New Society activists in the 1980s] as much as they do most anarchists today."[57] Generalizing from the experiences of Movement for a New Society, Cornell emphasizes that, "it is useful to distinguish between *organizational* and *social hierarchies*, and between *delegated* and *coercive authority*" in order to oppose domination and propose alternatives to the neoliberal competition state.[58] In other words, it is not egalitarianism in primitive freedom that is at stake but universalizing access to a decent quality of life and building the organizational means necessary to coerce political authority into reconfiguring the boundaries of political freedom to encompass these ends.

Mobilized in response to the race- and class-blind opportunity society promised by left neoliberalism and the increasingly naked appeals to racism and chauvinism accompanying a second generation of popular neocommunitarian politicking, advocates of a resurgent environmental justice movement have also sought to confront the maldistribution of economic and environmental harms in radical terms. Once limiting participation to consensus-seeking negotiations presided over by agencies such as the Environmental Protection Agency, the new environmental justice movement seeks to engage in political contest by extending longstanding civil rights demands that social harms must not disproportionately affect citizens by virtue of their race and class to encompass rights to a livable climate for all.[59] Longtime participants in the environmental justice movement recently urged a profound rethinking of aims amongst some of its members. This rethinking is important because the broader environmental justice movement has been deeply enmeshed within the left neoliberal governance regime and has by and large sought gains by also working with right popular neocommunitarian parties. Hence, sympathetic although stern critics Eric Swyngedouw and Nik Heynen argue that "the environmental justice movement speaks fundamentally to a liberal and, hence, distributional perspective on justice, in which justice is seen as Rawlsian fairness and associated with the allocation dynamics of environmental externalities." And, similarly sympathetic Ryan Holifield suggests that the environmental justice community's willingness to work through the EPA with left neoliberal and first-generation popular neocommunitarian administrations link that community to a "neoliberal project to *empower* and *build trust* in 'environmental justice communities'" and therefore not to "*redistribute* environmental risk more equitably."[60]

Extending this line of argument, Laura Pulido, Ellen Kohl, and Nicole-Marie Cotton describe how uncritical engagement by moderate environmental justice campaigners within but not against left neoliberal governance has delivered neither meaningful outcomes for constituencies nor durable structural or institutional change in the United States. Accepting that both left neoliberal and right popular neocommunitarianism offer two sides of the same coin, Pulido, Kohl, and Cotton urge environmental justice campaigners to ally more closely with radical groups:

> Environmental justice activists' reliance on state regulation has inhibited their ability to achieve their goals. . . . The fact that activists continually turn to the state and see it as the only option suggests the hegemony of the state in terms of creating social change: activists cannot readily identify paths outside of the framework offered by the state. . . . [However,] the state is not about to eliminate the necessary "sinks" that communities of color provide, for fear of both capital flight and the wrath of conservatives. Instead, the state gives lip-service to EJ but in fact does little to change the materiality of disproportionate pollution patterns. . . . Liberal groups will [and should] continue to work *with* the state, while [radical EJ groups can and should emerge to] *confront* it, perhaps through alliances with anarchist and/or anti-capitalist formations.[61]

In the context of hollow consensus building, movement co-option, and capture by industry characterizing several decades of involvement in left neoliberal and, sometimes but not always good-faith, popular neocommunitarian governance, these scholar-activists are urging environmental justice activists, whose main points of reference to date have been the achievements of the twentieth-century civil rights movement, into closer engagement with radical green-left groups.[62] This distinction between left neoliberal and popular neocommunitarian party machines and a radical green-left movement is of particular importance to environmental justice campaigners. For one such campaigner, Angela Park, at issue is not nature or carbon emissions, but "the way climate change affects communities, is embedded in social justice, impacts public health, and is intertwined with transportation and industrial facilities."[63] That is, these radicalized environmental justice campaigners recognize that successful political action depends on movement understandings of an indelible distinction between inside and outside, on engaging with structural conditions and institutional constraints and opportunities to advance constituency interests while actively building constituencies into politically powerful blocs that can oppose machine politics and protest injustice as the product of history and social context. This is because party-supported contemporary governance favors market-based and consensus-driven policy-making as does the mainstream green lobby for the most part. However, what Parks recognizes is that the mainstream organizations, "with media attention and significant funding, need the assistance and integration of the environmental justice movement in order to have an impact on many communities."[64]

Still others argue that awareness of the neocommunitarian dimensions of the governance turn places a demand on both radical movement activists and community organizers to distinguish between quietist civil privatism and green-left interests in undermining the social structures and institutions that facilitate the domination of some by others. James DeFilippis, Robert Fisher, and Eric Shragge argue that any effective response to left neoliberal ideology and the right neocommunitarianism it so closely resembles must recognize the real imbrication of the subpolitical ideal of community within a prevailing macropolitical structural and institutional order. For DeFilippis, Fisher, and Shragge, subpolitical ideas and community-centric practices—in particular, emphases on "cultural, consensual and parochial responses to concentrated power"—all too easily dovetail with the interests of a philanthropic donor class who champion "the primacy of the market and decentralization of the state."[65] In their view, left neoliberalism has undermined the capacity of radicals to organize political resistance to a well-organized and effective New Right. Echoing the new wave of post-Occupy radicals and environmental justice campaigners, DeFilippis and colleagues identify a "shrinking of perspectives" that has taken place amongst even radical activists who continue to work from the logical but outdated presumption that "challenging corporate exploitation and systems of official [and unofficial] apartheid" is essentially a "community-based effort."[66]

For DeFilippis, Fisher, and Shragge, the governance turn reduces "the frame of reference in community-based efforts to a focus on community in and of itself."[67] This community-based frame of reference is shown to be indistinguishable from that applied by left neoliberals in the philanthropic and responsible business sector and by New Right grassroots organizations such as the Christian Coalition, but for two important factors. First is the considerable amount of corporate funding available to left neoliberal "consensus organizing" groups that "assume that the inner-city poor and the rich and powerful have overlapping mutual interests and needs" and to "ultraconservative grassroots groups" such as the Christian Coalition and earlier Moral Majority.[68] Second is that New Right movements self-consciously mobilize community as part of a wider movement infrastructure. That is, they do so in contrast with prefigurative movements, which maintain a "narrow community focus" that "reinforces the legitimacy of the local community as the primary locus of action" and focal point for results.[69] Of course, radicals and even reformists can do little about the first factor but a great deal about the second. DeFilippis, Fisher, and Shragge identify three exemplary organizations—the Fifth Avenue Committee of Brooklyn, the Immigrant Workers Centers across the United States and Canada, and the now defunct Association of Community Organizations for Reform Now (ACORN). In their argument, it is only by utilizing a two-pronged inside-outside strategy that these organizations have been able to achieve measurable successes over several decades of rightward drift in American political culture. "Inside-outside" in this sense means recognizing larger state, national, and even global institutional contexts that communities fit within, and maybe also extending organizing efforts to those levels. They show that these three groups have expanded the remit of green-left politics by eschewing community-centric organizing and simultaneously rearticulating the premises of prefigurative activism in favor of an "emphasis on power, and conflict based organizing for social and economic justice."[70] The central factor in the successes of organizations that have grown in size, influence, and effectiveness since the financial crisis of 2007 has been their adoption of an inside-outside strategy and recognition of the limits of community-centric organizing and prefigurative activism. In order to become more effective, such organizations have eschewed ideas that are cut from the same cloth as that worn by both left neoliberals and popular neocommunitarians.

Other effective organizations that follow a similar approach to activism and organizing under the aegis of an inside-outside analysis are community-based social and environmental justice organizations that have emerged at the forefront of radical politics since the demise of Occupy in 2012, notably the Right to the City Alliance.[71] In the parlance of social movement studies, right to the city groups extend and elaborate a neo-Alinskyite" organizing model. As practiced by the Fifth Avenue Committee and Immigrant Workers' Centers such "progressive, neo-Alinsky organizing" eschews narrow community-centrism yet also avoids mid-twentieth-century activist-organizer Saul Alinsky's utilitarian assumption that community members are primarily motivated by (economic) rational self-interest. Organizing on the neo-Alinksy model thus maintains the original Alinsky model's emphasis on "conflict not consensus" while simultaneously provoking ideological awareness amongst a membership base. That is, the neo-Alinsky base-building model works from inside the political-economic system, usually through legal mobilization, while actively constructing and mobilizing action from outside, usually through public protest. In this sense, the neo-Alinsky model is premised on community organizing to achieve measurable goals and activism that creates "new political structures intended to replace existing political structures."[72]

In a parallel development within the labor movement, Jane McAlevey identifies a shift in the prevailing model of organizing since the 2008 crisis. For McAlevey, the failure of unions to increase membership significantly in the crisis's wake has been largely due to the predominance of a fee-for-service model of union membership. In her study, McAlevey faults Alinskyite unionism for embracing corporate campaigns and narrowly defined interest-based politics as decisions that led unions away from workers and the workplace and put them at odds with unorganized workers and the community. In contrast, she highlights the moderate successes of unions allied with the Change to Win minority coalition within the movement. In her view, Change to Win unions are relatively successful because they reinvigorate pre-Alinskyite strategies of supporting organic worker leaders in the shop and anchoring campaigns in the 'whole worker,' a person embedded in a range of social relationships in the workplace and in the community.[73]

In part responding to such developments, some neo-anarchist activists have adapted the inside-outside model to environmental justice campaigning along lines set out by Right to the City. Many associated with Right to the City were deeply if not uncritically engaged in the Occupy protests.[74] An illustrative example here is the Nobody Leaves Mid-Hudson organization. Nobody Leaves Mid-Hudson is headquartered in Poughkeepsie, New York, and allied with several similar post-Occupy base-building environmental and social justice movements nationally, including Right to the City's Homes for All campaign; they are also allied with other post-Occupy environmental justice groups across the United States, including Southwest Workers Union, Asian Pacific Environmental Network, and Kentuckians for the Commonwealth. Indeed, Nobody Leaves Mid-Hudson was founded by long-term community activists, many of whom had been foreclosed on in the past. These activists came together and launched the group in late 2011 as former participants in Occupy and Poughkeepsie's anti-foreclosure working group.[75] Building on Right to the City's principled assertion that nobody should be forced to leave their local community due to economic distress, the group calls for the integration of household energy issues into environmental justice organizing. Elaborating the inside-outside strategy, this group has developed an effective model of organizing "which combines individual case advocacy; local base-building; direct action and policy advocacy at local, regional, and state levels; and participation in broader environmental alliances,"[76] notably the New York Energy Democracy Alliance.[77] It is worth quoting at length from Nobody Leaves Mid-Hudson's 2016 report,

> Just utilities policies that empower rather than immiserate poor communities and communities of color are a critical part of environmental justice and energy democracy. Incorporating affordability and other policies can greatly strengthen the environmental movement. For low-income communities not facing severe local environmental threats, the household energy crisis is more immediately felt than the problem of unhealthy and environmentally destructive fossil fuels. However, even in severely environmentally threatened communities, energy unaffordability and inaccessibility are experienced as critical hardships. Therefore the environmental movement must embrace household energy security as a primary mobilizing issue. The private organization of energy distribution raises the issue of corporate rather than public and community control of energy, though in a different realm from production. Corporate-controlled distribution works against democratic discussion of energy alternatives and community control of clean distributed energy. Without adequate consumption of energy it is impossible for households to maintain healthy every-day lives and community participation, including participation in environmental movements. The severity and scope of the energy affordability crisis demand that the environmental justice movement incorporate energy utility affordability for the movement to be widely relevant to low-income people and people of color and to open significantly greater space for low-income communities' leadership in the movement. By addressing this urgent issue of consumption, longer-term participation in struggles for cleaner and community-controlled energy becomes possible for low-income people of color in frontline communities.[78]

This kind of base-building organizing is of course not new. However, Nobody Leaves Mid-Hudson combines sophisticated structural analysis of what is often taken to be a secondary or consumer issue with awareness that it is not the prefigurative practices of activists but the entire base within which a political movement sits that matters to confronting domination. This combination situates organizations such as this one at the forefront of green-left renewal. Put differently,

> energy utilities are a strategic issue, not simply for winning energy utilities justice. They are also strategic for building the power of low-income people and people of color to win a range of specific immediate material gains and broader racial, gender, economic, and environmental justice. They can strengthen the housing justice movement, link the housing and environmental justice movements, and bring low-income communities of color into leadership in the environmental justice movement. Campaigns around energy utilities have strategic advantages that flow from the scope and severity of the issues, the geographic organization of energy utilities, and the role of state institutions.[79]

Clearly then, the structural moralizing purity of the vision for another possible world takes a back seat to clear-eyed analysis and strategic partnership. What is central here is not prefiguring in the here and now some abstract and inalienably true human good or any infinitely demanding ethics. The organizational task is understood as one of building a constituency that knows who benefits from which institutions and developing the kinds of solidary action required to alter the rules under which such institutions operate. Nobody Leaves Mid-Hudson is addressing specific problems that are consequences of a particular historical, political, and therefore economic context, which have unfolded at a distinct scale in different and sometimes contradictory ways and have exerted impacts that are experienced by particular groups and individuals. The externalization thesis that underpinned the resurgence of radical green-left activism in the late 1990s has come full circle. The effort here is not to compel the internalization of externalities but to make explicit the terms on which the political and economic conditions that encompass housing, utilities, and indeed, environmental justice concerns more broadly, have facilitated the domination of some by others, and to alter this through political action.

A new political sensibility is thus emerging. As seen in the post-Occupy soul-searching by some neo-anarchists, the calls for a two-pronged strategy by environmental justice campaigners, the examples offered by DeFilippis, Fisher, and Shragge, the illustrative example provided by Nobody Leaves Mid-Hudson and allies, the efforts of activists within the Democratic Party such as Working Families, the discourse presented in new publications such as *Jacobin*, the work of the Democratic Socialists of America, labor-green coalitions, and the feminist and LGBTQI movement, a sensibility that eschews the prioritization of moralizing in favor of engagement in politics has taken hold. Grounded by direct experience of the failures of the Occupy moment, these and myriad other groups seek to operationalize a new politics that emerges not from within activists' and organizers' own moral choices to live ethically but from a base in particular communities and in relation to distinct issues: exposure to environmental pollution amongst minorities, a living wage for workers, civil rights for historically marginalized groups, and a just and fair transition to sustainability for the poor, for example. That is, they are mobilizing in ways that expose the externalization of social and environmental harms as an essential consequence of the terms on which currently legitimate responses to the first political question allow the domination of some by others. Critique of and confrontation with political authority are embraced. Constituencies are built around shared recognition of the need for an effective movement that is evaluated in terms of its purchase upon rationalized authority. Whether emerging from reflexive self-criticism by post-Occupy neo-anarchist scholar-activists, Democratic Party activists in the wake of the 2016 presidential election, the environmental justice movement, or radical community organizers with first-hand experience of the failures of Occupy, contemporary radicals and reformists are moving towards the building of properly political projects that aim to undermine structural and institutional barriers to social and environmental justice.

"Meaning" in the sense that Geuss implies has been refound, albeit in truncated and negative form. The new wave of radicals and reformists remains cognizant of the fact that justice requires significant and effective blocs of citizens and that what such blocs can achieve remains limited in the current context. The New Right's efforts to defend untrammeled private property rights and a narrow vision of justice as wealth maximization have put utopian dreams on the backburner. Moreover, the climate change that was in the 1970s opposed as a future possible is now a reality, and the science shows that even the most optimistic left neoliberal plans for action might well prove insufficient to mitigate its effects, let alone halt it. This is no reason to give up on the radical project. It does, however, imply that critique oriented to achieving justice as the expansion of political freedom—the artificially sanctioned status of the citizen as autonomous in relation to others and the state—may be delimited to negative demands. Such demands are negative in the sense that radicals and indeed reformists promoting them recognize that those who benefit the least bear little responsibility for positive plans to achieve sustainability. The demand for justice in the Anthropocene age of ecology may be limited to disrupting whatever plans do get made in order to ensure that they do not disadvantage the poor or historically marginalized.

Conclusion

Authority and Meaning

Forced by circumstance to reflect deeply on almost half a century of commitment to an array of ideas and practices that have achieved little, radicals and reformists alike have experienced a crisis of meaning. Since the evanescent moment of Occupy and, more so, of the New Right's mobilization in the lead up to and following the Republican presidential victory of 2016, a surge of critical reflexivity has taken root amongst a new wave of scholars and activists. Consolidated amongst radicals has been the shared recognition that strongly held moral convictions, be they informed by earth science or the social sciences and humanities, do not necessarily translate seamlessly into effective political agency. Following a long decade of Right political and cultural successes, the financial crisis that erupted in 2007 had similarly provided an opportunity to renew the reformist agenda. However, Democratic Party leaders proved more willing to favor the interests of financial marketeers over those of citizens harmed by predatory lending or rendered jobless or underemployed by the crisis. The opportunity to recant left neoliberalism was missed. The capitulation of candidate Hillary Clinton's campaign to second-generation populist neocommunitarian Donald J. Trump all but sealed the Right's long worked for ascent. Although losing the popular vote by close to three million ballots, Trump won the presidency, the Republican Party won a majority in Congress, and, together, the power to nominate and confirm appointments to the Supreme Court.

For many concerned with the rise of the Right and the so-called alt-Right, the election result brought into sharp relief the effectiveness and organization with which people on the right had since the 1970s been pursuing their aims.[1] The Right had consolidated itself as an influential and effective political movement around two discursive themes. First, constant reiteration of a longstanding rhetorical device—the moralized politics of fear—helped activists and organizers to wed the interests of the wealthy to the grievances of working and lower middle-class constituencies of "ethnic" white northerners and white southerners bearing the brunt of globalization, as well as the "revanchist middle strata" of precariously situated yet aspirational suburban and exurban denizens of the new economy. Second, central to this moralized politics of fear were portrayals of the rump welfare state, scientific knowledge of human-induced climate change, and social science knowledge of the structural sources of economic inequality as the illegitimate products of a combination of elite hubris and individuals' moral failures. Such hubris had indiscriminately offered handouts to the undeserving and appeasement to "the loony Left" while handicapping the efforts of the righteous, and so the national community, to prosper.

The grounds for such mobilization were reactionary hierarchical and particularist beliefs that the patrician conservative Right's acquiescence to radical and reformist demands was not only a quixotic endeavor but also harmful to the interests of the deserving. Aimed initially in the early 1970s at reining in what was thought to be the leftism inherent in cultural modernism, New Right intellectuals worked to overcome Hayek's paradox and by the early 1990s had refashioned themselves champions of modernism. By the late 1990s, the New Right had seemingly convinced a significant number of citizens that any collectively organized effort to shape the contours of political freedom in terms other than those dictated by humility in the face of Darwinian struggle for wealth maximization would impinge upon the freedoms of struggling "real Americans" as well as those who had succeeded in that struggle: financiers, rentiers, entrepreneurs, owner-managers, and corporations themselves as "legal persons."

Throughout this book, I have looked to different perspectives on this confluence of events. The Foucauldian approach used by Luke, and in a related way Brown, has exposed how the creep of biopolitical power has subsumed the once emancipatory ideal of individual freedom to the relentless logics of alternating ecological and economic iterations of governmentality. Meanwhile for interpreter of Luhmann's social systems theory Blühdorn, Enlightenment's contradictory commitments to idealized freedom and instrumental rationality have brought about a situation in which a system premised on instrumental rationality has internalized demands mobilized by that ideal. In cultural historical terms and specifically focusing on the United States experience, Steve Fraser regards the collapse of the modern Left as a product of the displacement of class awareness by 1960s-inflected libertarian individualism. In a similar way, Frank ties this collapse to the increased influence on both mainstream and radical politics of affluent technocrats, themselves clinging to ideologically libertarian ideas that first circulated amongst the New Left in the 1960s. On a different plane, intellectual historian Rodgers identifies a process of displacement in public debate, whereby academic efforts to elaborate thin metaphors of choice, agency, performance, and desire have come to replace concrete demands for explicit kinds of justice based on thick metaphors of social status, economic circumstance, and institutional remit. And, I have examined work by political and policy historians and historians of ideas Fox Piven, Philips-Fein, Mirowski and Plehwe, Dardot and Laval, Maclean, and Amadae, who trace at least some of the sources of this displacement to the often privately organized and funded policy recommendations, social scientific, and jurisprudential arguments developed by an array of wittingly or unwittingly right-leaning intellectuals, academics, business people, lawyers, and judges whose actions helped to overturn many of the achievements of radical and reformist agency aimed at advancing the Enlightenment project.

Informed by this work, and mobilized by Geuss's essay, I have used Adorno's prescient modernist sensitivity to the weaknesses inherent in the radical Enlightenment project as a guide to interpreting a situation in which "people on the left find increasingly that they have lost faith in the traditional diagnosis or in some part of the traditional recommended therapy" for challenging domination as unfreedom. Adorno's thought is helpful insofar as he remained acutely aware that the radical Enlightenment project depends on negative, nonidentity thinking, which involves considered rather than absolute criticism of identity thinking alongside absolute opposition to what he called false nonidentity thinking. Until the late twentieth century, meaning for those like Adorno, whom Geuss terms people on the left, had been cast in modern terms. Adorno wrote at a time when the United States was one of two preeminent modern political-economic and cultural powers. In that period, both radicals and reformists had found meaning in citizen movements oriented to coercing or capturing the authority held by those controlling the apparatus of state. By extension, and as Adorno, with Horkheimer, had recognized, meaning was grounded in the Enlightenment aspiration to exert rational mastery over nature in order to expand freedom.

The loss of meaning became apparent to thinkers such as Adorno as he recognized that unlike premodern authority, authority oriented by the task of exerting rational mastery over nature for human ends was bound to concede something to negative critique. More so, following Hegel and Marx, Adorno saw that critique was itself a product of Enlightenment reason. As the Soviet experience of state communism demonstrated most vividly, an authoritative system that failed to concede something to negative critique would atrophy and eventually collapse under the weight of internal contradictions. Yet, as Adorno seems to have understood, and as the experience of freedom under the liberal- and social-democratic systems of the mid-twentieth century long upturn demonstrated, critique itself must ultimately remain negative. Albeit with considerable bitterness and regret, Adorno came to recognize that the dualistic ontological and epistemological-ethical presumptions of Enlightenment that had informed and given meaning to modern critique were unsatisfactory yet necessary conditions: "The whole is the false." Arguably the source of Adorno's regret is his realization that the experience of freedom always involves identifying with the system against which negative critique has pushed. With this realization of the condition of "entanglement" in hand, only the philosopher's infinitesimal freedom remains.

For better and worse, Adorno achieved such clarity through his exhaustive commitment to modernism, a sensibility that accepts enlightened secularization but rejects the dualistic modern desire for rational mastery of nature. Modernism for him involves embracing reasoned self-direction in order to make one's home within the whole of reality, the universe of natural forces as transformed by human action. Hence, Adorno opposes the unreflective, and so positive, identity thinking that he attributes to the moderns with an always-negative commitment to truthfulness. This commitment to truthfulness in the face of the sublime, of awe-inspiring reality, is for Adorno what sustains his modernist nonidentity thinking. Yet, his modernism entails only this negative commitment to critique and goes no further. Following Nietzsche's critique of "bourgeois" modern subjectivity, Adorno's modernism is a reflective *philosophical* project that aims to sustain autonomy amidst the modern system: "Wrong life cannot be lived rightly." For Adorno, the modern Left had, since Marx, if not since Rousseau, problematically embraced the political, positive task of making things better. What Adorno's modernism allowed him to show was that, convened as a political project, the modern sensibility readily morphs into identity thinking, into support for the pathological instrumental rationality of the utilitarian calculating ego. However, he also recognized that efforts to channel modernist nonidentity thinking into any kind of positive collective or, more pointedly, communitarian or nationalistic project would bring even worse. This realization was the source of both his melancholy ambivalence about rationalized authority and his fear of the "horror of the diffuse," of false nonidentity thinking. As Williams recognizes on the basis of a similarly if vindicatory rather than critical Nietzschean logic, both identitarian "enactment" and falsely nonidentitarian "structural" orientations to practical reasoning about the relationship between means and ends commit those adopting them to positive truth as essence. In the realist terms favored by Williams, to embrace positive truth as essence as a guide to action is to abandon the space for critique that is opened by rationalized authority and, therefore, to undermine it.

Engaging Williams's work on the politics of legitimation and justification under rationalized authority, I have argued that Adorno's tripartite distinction between a modern sensibility grounded in identity thinking and two distinct iterations of the modernist sensibility, false nonidentity thinking and nonidentity thinking, can be observed in action at the level of ideology. In this way, I have looked at political philosophies and political theories and the philosophers and theorists who recommend them as issuing politicized statements that seek to explain and inform actors engaged in political conflict. Such actors mobilize around one of the three practical rationales, using them as justificatory schemas for the legitimation or critique of authority and those holding it. These rationales are grounded by longstanding commitments to justice based on the utilitarian calculus, some inalienable truth, and an agnostic commitment to expanding political freedom as a condition of civil peace. For deeply entrenched historical reasons, one longstanding and highly influential rationale in the United States has been that Williams call enactment moralizing. This is the panoptical utilitarianism that seeks to shape the behavior of individuals in order that they *enact* the requirements of a maximally efficient system. This is the justification for authority as wealth maximization, which regards the calculating egoist, the self-interested monad encountering a wild, empty frontier-like natural order, as the worthiest contributor to the prevailing order. As I have argued in relation to the green-left modernist movements that first arose in the 1960s and 1970s, another longstanding if less influential justification for authority has used ideas and actions to mobilize support for some inalienable truth, such as the individual rights or ecological harmony or equilibrium that should *structure* authority in the image of the worthiest actor. Indeed, radicals and reformists have often invoked such inalienable truths in counterpoint to egoistic utilitarianism. Yet, from the perspective afforded by Adorno's "detached observer," when put into practice the two historically salient justifications combine to effect a peculiar situation. This, Williams saw as confining the study of politics and the utterance of political statements by philosophers and theorists in the United States to an

> intense moralism which is predictably matched by the concentration of American political science on the coordination of private or group interests: a division of labour which is replicated institutionally, between the "politics" of Congress and the principled arguments of the Supreme Court. . . . That view of the practice of politics, and the moralistic view of political theory, are made for each other. They represent a Manichean dualism of soul and body, high-mindedness and the pork barrel, and the existence of each helps to explain how anyone could have accepted the other.[2]

At issue in the cases of enactment and structural moralizing is that they sustain mutually exclusive understandings of what it is to justify the exercise of power under authority. I have argued that the two historically prevailing forms of practical reasoning confine advocacy to the expression of one of two different forms of the naturalistic fallacy. On the one hand, advocacy might presume that the malignant, raw power necessary for success in Darwinian struggle is all that matters to the exercise of authority, and presume that political institutions and the individuals engaging with them can operate on principles that benefit the interests of the strongest, most deserving, and so most rightfully powerful. This is advocacy on the model of enactment moralizing and I believe gives practical expression to what Adorno saw as philosophical identity thinking. On the other hand, advocacy might presume that the benign, mutually beneficial, and cooperative goods of which relatively homogeneous communities of moral actors are capable can also potentially be exhaustively embodied in amoral politically structured institutions. This is advocacy on the model of structural moralizing and I believe gives expression to what Adorno saw as regressive philosophical false nonidentity thinking.

The third form of practical reasoning, agnostic public welfarism oriented to asserting claims against others under political authority, did exert a strong influence on the political climate of the mid-twentieth-century United States. However, since the mid-1970s, as a motive for action, public welfarism has declined in influence relative to egoistic utilitarianism and commitments to the pure inalienable truth. The form of practical reasoning which supports begrudging involvement in solidary politics oriented to forcing some form of institutionalized constraint on both state interests in security and peace and market interests in profit is, in light of the considerable caveats that I have presented here, what I believe Adorno's concept of nonidentity thinking makes visible for a critical political theory. The most important caveat that I recognize is that Adorno's efforts were philosophical and as such prioritized as axiomatic the theory of freedom. As Adorno himself made clear, his philosophy offered no grounds for a collective political project. Yet, understood in the context of his extra-philosophical "real life" activity in support of public education, social democracy, and what has recently been recognized as his advocacy of democratic citizenship, his philosophy of survival in modernity offers a superlative analytic perspective upon the critique of injustice in the Anthropocene "age of ecology." When interpreted in light of Williams's realist account of the binding of political authority to respond to the first political question in ways that consider appeals not only to raw power but also to principle, the tripartite structure of Adorno's philosophy facilitates a degree of detachment from critical praxis. In this way, I have moved away from the pure philosophical theory of freedom that guided his critical theory and toward an understanding of freedom as a "ratio concept" that is inexorably related to justice, authority, and legitimation.[3]

Hence, I have examined Adorno's philosophical arguments, which he develops from the normatively problematic axiom of mimetic reconciliation, as historically situated responses to conditions in the mid-twentieth century, when he and Horkheimer first identified and subjected to critique the myth of Enlightenment reason. I have understood Adorno's work to not only interpret but to express some of the tensions between those who sought power in the United States and those who opposed it in the mid-twentieth century. Yet, his work also offers insight into developments since the late 1960s, the historical period in which the myth of Enlightenment reason lost its power to enchant. As Radkau shows, it has become increasingly clear that the ontological certainty that humanity can exert rational mastery over nature has given way to a new ecological view of humanity as but one participant in the whole encompassing ecosphere. This is to say, in conditions that Adorno and his contemporaries could not have foreseen, amidst a political economy that has come to depend for its reproduction on the modernist sensibility that was once the preserve of philosophical radicals such as him. And, more importantly, amidst a political economy that seems increasingly to facilitate authoritarian iterations of the malignant naturalistic fallacy. Belligerently communitarian and nationalistic, second-generation popular neocommunitarians such as Donald Trump, as well as many of his followers, seem to espouse ideas that seek to subsume all to identify with the logic of Darwinian struggle. This is ideology on the model of enactment moralizing insofar as its authoritarianism is naturalistic, holistic, hierarchical, and particularistic, even as many of its advocates deny the reality or pressing nature of human-induced climate change and champion their love of primitive freedom.

This is important because it is arguably the case that in the mid-twentieth century, as scientists first definitively established that the "advanced industrial" societies were inducing climate change, the workers,' women's, civil rights, and Green movements made significant gains. With hindsight, this period marked twilight for the myth of Enlightenment reason. This was also a period in which radical and reformist movements accepted that organized citizenly power—exerted through labor unions, civic associations, and similarly organized machinic solidary movements, including the reformist Democratic party—was a sufficient and necessary condition for achieving such gains. Such protest was committed to truthfulness in the face of the experience of injustice as Adorno had understood it. Wittingly or not, the result was that the protesters acted in ways that coerced those holding authority to respond differently to the first political question. These often violently suppressed movements sought varied, often opposed ends, especially when considering opposition to civil rights amongst elements of the labor movement. The responses of those in authority, however, were of a kind that aimed to maintain civil peace. Such a peace simply appeared relatively more just to those with no choice other than to subject themselves to political authority, but who nevertheless were ready and able to disrupt the peace through strikes, sit-ins, boycotts, voter and labor union registration drives, draft resistance, and, albeit unsuccessfully, working to dismantle and remake the Democratic Party. These actions and demands expanded the contours of political freedom. With hindsight, it may be surmised that the path to some semblance of justice as freedom under rationalized authority lay in efforts to extend political freedom at the expense of those better able to influence authority, and not in efforts to remake authority in the image of some benign inalienable truth or malignant Darwinian fantasy. Such a critical political realist perspective becomes even more pressing when considering that the ascendant Right's ultimate goal appears to be that of dissolving artificial political authority, and replacing it with moral authority that is ultimately oriented to upholding raw power: that is, of reviving an understanding of authority as the capacity of elites, the survivors who prosper in the raw natural struggle, to mimetically express such cosmological forces.

Coda: Agonism or Agnosticism?

Before ending, I briefly consider one highly influential and current critical effort to guide action and re-form radical and reformist thought and practice along similarly realist lines. Political theories of "democratic agonism" challenge the exertion of postpolitical authority aimed at shaping individuals in the image of the egoistic utilitarian calculus. One advocate of such agonism is Chantal Mouffe, who has sought to remobilize the radical project in postpolitical conditions and along realist lines.[4] As a critical response to the de-democratizing impacts of the "governance turn," Mouffe's agonism makes explicit, rather than seeks to reconcile, tensions between the holders of different interests. Her work achieves this by delineating clearly between historical, structural, and ideological conditions that allow the domination of some by others and the options for political agency to which those suffering may resort.

Mouffe's agonistic theory conceptualizes *politics* as the realm of the essential, of necessity, order, control, conventions, institutions, and constraints on human agency and interactions. She conceptualizes *the political* as the realm of anti-essentialist, antagonistic forms of agency, agonistic respect for difference, disequilibrium, and, in effect, being as becoming. The sensibility that Mouffe associates with the political thereby closely resembles what I have described, after Adorno, as modernism. Mouffe draws this distinction using the same Heideggerian strand of postmodern theory that was taken up by green-left modernist movements in the 1990s, albeit tempered by the work of Heidegger's friend, the jurist Carl Schmitt.[5] In her view, modernity is defined by the inexorable tension between politics and the political. Reconciliation of the tension is achieved through "hegemonic projects." As a consequence of liberal democracy's particular history, which includes Enlightenment values, hegemonic projects justify the interests, morality, and ethics of those holding power at the core of the state-market nexus. The hegemonic project of modernity is for Mouffe that of Enlightenment, which as politics fetishizes essentialist notions of reason, universalism, and individualism. In contrast, the political—the "axiological principle" that always and everywhere opposes politics—confronts Enlightenment with its antitheses: disorder, contingency, unconventional ways and means, spontaneous actions, unconstrained creativity, and so on.

Unfortunately, Mouffe uses this distinction to drive a normative argument from ontological truth. The realm of the political is necessarily the site from which radical movements do and should oppose domination as unfreedom. Freedom is in this view pre-theoretically predefined as what allegedly goes on in the realm of the political: spontaneous, anti-essentialist agency, celebratory pluralism, authentic creativity, agonistic respect for difference, disequilibrium, and genuine being. It follows that, after Heidegger, the hegemonic project of modernity, Enlightenment, seeks to annul these. For Mouffe there is a tendency for radicals to ignore what she regards as the essential tension between politics—the universalistic, rationalizing, and individuating impulse of Enlightenment's hegemonic project—and the normatively desirable political—the particularistic, irrational, and expressive human capacity for jouissance. The result is that many, including the most vocal of the radical green-left modernist revivalists of the 1990s, regard critique as necessarily involving withdrawal from politics. In Mouffe's acute critique, this means that, while "the Left shows little interest, Right-wing and authoritarian groups [have been] only too happy to take over the state."[6]

So far, it seems fair to say with Mouffe that a key failing of the prefigurative communitarianism that I have ascribed to the scholar-activists of the green-left modernist revival that began in the wake of Radkau's second great eco-boom of the 1990s was their tendency to mistake withdrawal for engagement. However, Mouffe does not call for radicals to engage the state by coercing those holding power within it to expand the contours of political freedom. Rather, she makes a somewhat counterproductive argument. She calls for radicals to embrace an abstraction: the modernist anti-essentialism essential to the political. In

Mouffe's view, radicals cannot seek to replicate politics but rather should seek to "engage with the state" through deployment of

> counter-hegemonic practices, by contrast [these], do not eliminate differences. . . . They are . . . an "ensemble of differences," all coming together, only at a given moment, against a common adversary. Such as when . . . environmentalists, feminists, anti-racists and others come together to challenge dominant models of development and progress.[7]

The issue that Mouffe's thesis raises is in fact identical to that raised as a practical problem by the reflexive radical scholar-activists of the post-Occupy new wave. What Epstein, Blair Taylor, Dixon, Cornell, DeFillipis, Fisher, Shragge, and others seek to reevaluate is the normative value to the radical project of precisely what Mouffe defines as the political. The problem that Mouffe's agonistic theory confronts is twofold. First, once an "ensemble of difference" has disrupted the political at the level of a hegemonic discursive intervention, the problem of maintaining the political power sufficient to maintain or further alter the contours of a newly expanded political freedom remains. Second, agonism presumes that the constituencies that radical and reformist movements give voice to share equally in scholar-activists' enthusiasm for anti-essentialism. As Lois McNay cogently observes, the agonistic turn is, theoretically speaking, weightless.[8] Mouffe presumes that the jouissance of the political can be lastingly formulated as a basis for the solidarity necessary to coerce political authority to respond differently to the first political question.

Mouffe's calls for unrelenting agonism do however resemble Adorno's unrelenting demands for truthfulness in the face of the sublime. The difference is that Adorno was well aware that such demands were aesthetic and offered philosophical insight. Modernism offers no guideline for collective political agency. This is because modernism informs a very particular attitude towards authority. The modernist sensibility embodies a prepolitical truth concerning the exercise of subjectivity in or perhaps, against modernity. The problem is that once modernism is accepted as a normative guideline for authentic collective agency, all that remains are questions of moral authority. As Adorno recognized, these all too easily devolve to assertions of raw power. Approaching injustice by settling such ontological questions first and setting out a normative political program oriented to justice based on such settlement is problematic for political critique. When bought into contact with power, action informed solely by structural moralizing from inalienable principle readily "flips" into coercive enactment moralizing. Understood in context, what Mouffe defines as the "axiological principle" of the political that is always and everywhere opposed to politics—principled disorder, contingency, unlimited diversity of lifeways and viewpoints, unconventional ways and means, spontaneous actions, unconstrained creativity, and other such expressions of modernism—is in reality more akin to the neoliberal conception of agency in a society where wealth maximization defines justice than any kind of political solidarity that she otherwise strongly supports as the foundation of social democracy.

Reproducing society undoubtedly involves ontological commitments. Yet, understood in context, the actions that reproduce such commitments are also dependent on artificial institutions—states, markets, and inter-, non-, and increasingly anti-state organizations, for example—as well as informal structuring frameworks—such as ideologies, norms, beliefs, and values. The modernist capacity to experience the jouissance of what is essentially primitive freedom is a capacity that results from one's status as the beneficiary of a very particular set of historical circumstances. Once posing a radical threat to hegemonic modern political culture, modernism is today used to legitimate a host of institutional arrangements that make it more likely that many suffer while some prosper. Not least amongst these are institutional arrangements that prioritize an idealized freedom that can only ever be exercised by the few. In the early twenty-first century, the naturalistic ideal of freedom that was incorporated from critical theory by deep and social ecologists and communitarians in the 1970s and that did clearly confront the centrally planned and organized despoliation of the entire ecosphere and the industrial system of mass domination has emerged as a source of injustice. Today, such a thin abstraction is all too readily amenable to the exercise of raw power and so primitive freedom. And, primitive freedom is always the freedom of the strong. In Geuss's oft-cited formulation, prevailing ideas and beliefs about the world, about the "true" constitution of it, are

> usually dead politics: the hand of a victor in some past conflict reaching out to try to extend its grip to the present and the future. There is nothing inherently wrong with this. Our past is an essential part of what we are, which we ignore at our peril. We could not leave it behind, even if we wished to do so.[9]

In the United States in the early twenty-first century, Geuss's "hand of the victor" may just be that of a once radical but now institutionalized cultural modernism: naturalistic, holistic, particularist, and, wittingly or not, hierarchical and voluntaristic.

As the modernist Adorno saw, truthfulness can only emerge from the injustice experienced by those who would otherwise be freely self-directing individuals, but for the positive commitment to truth as essence that the modern system of social reproduction demands:

> The power of the *status quo* puts up the façades into which our consciousness crashes. It must seek to crash through them. . . . Surviving in such resistance is the speculative moment: what will not have its law prescribed for it by given facts transcends them even in the closest contact with the objects, and in repudiating a sacrosanct transcendence. Where the thought transcends the bonds it tied in resistance—there is its freedom. Freedom follows the subject's urge to express itself. The need to lend a voice to suffering is a condition of all truth. For suffering is objectivity that weighs upon the subject; its most subjective experience, its expression, is objectively conveyed.[10]

What Adorno could not have foreseen is that the ontological commitment and ethical priorities that informed radical modernism in the context of United States political culture in the mid-twentieth century would come to serve very different ends in the twenty-first century. Once radical, the ontological commitment to holism and concomitant ethic of moral voluntarism today serve as tools for the revival of the hard Right and for an all but exhausted left neoliberalism. Ideologists of the New or alt-Right justify action as morally worthy when expressing Darwinian principles, while those still adhering to left neoliberalism dream of a benign and global *natural capitalism*. Both prioritize the beneficial consequences not of an artificially constituted civil peace in which what counts as public welfare might be contested as an illegitimate reduction of political freedom but as the aggregation of the choices made by moral volunteers. The employment of holism and voluntarism by left neoliberals and by popular neocommunitarians in this view represent plays in a long game of ideological positioning and repositioning oriented by modernist values. Through this long game, those who benefit most from ostensibly free markets have been able to all but undermine the legitimacy of the political authority that once, incompletely and begrudgingly, conceded to accept nonmarket values and even uphold deontological institutions that would in some small ways respond to those who benefit least. What Adorno, together with Horkheimer, had regarded as the premodern West's reenchanting myth—the mimetic self-representation by elites of themselves as expressing, imitating, or replicating the natural cosmological order—has been revivified.

In place of deontological institutional constraints on the exercise of power under authority have emerged zero-sum evaluation criteria, assessed in the same terms as are natural cosmological forces, such as the laws of entropy or species selection. In place of deontological institutions—upheld as such by historically constituted yet today largely disabled solidary blocs of citizens—has been established a model of authority that refers to unfalsifiable ontological truths. Postpolitical moral authority, as exercised most clearly by legal pluralists and the doctrine of law and economics, presumes that value is an ontological feature of the world that exists prior to history and political contest. A neo-Enlightenment mythology that portrays complex natural equilibria attained through the facilitation of individuals' "free" moral choices to make their own homes within modernity undergirds postpolitical moral authority. In Adornian terms, hierarchy and particularism, and the violence and cruelty of "blind nature," have been bought back to life. Mimetic regression represents the negative perfection that Adorno feared would accompany the demise of the identitarian Enlightenment rationality toward which he was so ambivalent. The "return" to mimesis disqualifies the "infinitesimal freedom" that secular reason affords, the solidarity that democratic citizenship offers, and the egalitarianism that universalism provides for but, of course, does not instantiate in the absence of negative critique. Such is the problem that Adorno foreordained when juxtaposing the pathologies of identity thinking—of a truncated reason that accedes to and even "progresses" by virtue of the immanent critique posed by nonidentity thinking—against the horrors that would accompany the resurgence of false nonidentity thinking.

As powerful and wealthy actors work to "make America great again" with callous disregard for the consequences, handwringing faux-critical accounts of our responsibility for undoing the deleterious effects of the Anthropocene merely replace radical and reformist politics with irenic belief in the power of well-meaning consensus. The holistic and voluntarist reenchanting myth of neo-Enlightenment in fact provides an apolitical Get Out of Jail Free card for both radical political ontologists and left neoliberals: over the long run, evermore well-informed consumers or right-thinking community members will choose the good; a rising tide of sustainability and harmonious Simply Living that will lift all boats. For populist neocommunitarians, this same mythology serves as a straw man for the moralized politics of fear: as "do-good" businesses and policymakers adopt unnecessary social and environmental policies in pandering to urban liberals, average Americans will be priced out of the markets for goods and services that are their rightful gains in a nation pursuing its manifest destiny.

In a critical lens, then, justice cannot today be fruitfully conceived as the expression of natural ontological "whole" properties of the world. Political actors, including scholars, simultaneously inhabit two different realms. One is the *natural* realm of ontology—wherein the universe has meaning in itself, as an ontological feature; that is, offers "doctrinally pure" moral guidance, be it malignant or benign in temper. When seen from within the ontological realm, all social relations are set along a moral continuum that is the natural order of things. Influential actors today are often willing and able to exploit "ontology" as a regulative idea. Competing moral visions for a viciously identitarian Darwinian or a naively benign and falsely nonidentitarian "eco-friendly" global market order share in common absolutist commitment to a world that has meaning in itself, as an ontological feature. The other is the *artificial* realm of politics—in which humans act collectively over time to reproduce a world that is held in common but that is not necessarily just. Such a realm is also part and parcel of "reality" but operates on often contradictory yet rationalized principles that might derive ultimately from sentient beings' capacities for imagination and delusion. Philosophical reflection can only help one to endure a wrong world. For those concerned to act, the partial and incomplete justice that is obtainable as political freedom under artificially rationalized authority might regrettably be the best focus for radicals in the so-called Anthropocene.

Notes

Introduction

1. For Marxist accounts, see work by Adolph Reed Jr. and Walter Benn Michaels, who link the current crisis of radicalism to the displacement of socio-economic class by cultural identity politics, and for a liberal view, see work by Mark Lilla, who links the crisis of reformism to the abandonment of national, state, and local organizational politics in favor of a highly intellectualized campus politics. See Reed, "Marx, Race, and Neoliberalism," *New Labor Forum* 22, no. 1 (2013): 49–57; Michaels, *The Trouble with Diversity: How We Learned to Love Identity and Ignore Inequality* (New York: Holt, 2006); and Lilla, *The Once and Future Liberal: After Identity Politics* (New York: Harper, 2017).

On the radical Enlightenment project, see Margaret C. Jacob, *The Radical Enlightenment: Pantheists, Freemasons, and Republicans* (Boston: Allen and Unwin, 1981); and Jonathan I. Israel, *The Radical Enlightenment: Philosophy and the Making of Modernity, 1650–1750* (Oxford: Oxford University Press, 2002).

On the enduring influence of critical theory in the United States, see Patricia Mooney Nickel, ed. *North American Critical Theory after Postmodernism: Contemporary Dialogues* (New York: Palgrave Macmillan, 2012).

Raymond Geuss explains that his argument, published as a book chapter titled "The Loss of Meaning on the Left" in 2014, was initially presented as a conference paper in 2008. See Geuss, *A World without Why* (Princeton, NJ: Princeton University Press, 2014), 111, 247n.

2. For a brief account of how such a threat may be manifesting in the United States and Western Europe, see Jan-Werner Müller, *What Is Populism?* (Philadelphia: University of Pennsylvania Press, 2016).

Chapter 1

1. For an overview of the now extensive realist literature, see, for example, Enzo Rossi and Matt Sleat, "Realism in Normative Political Theory," *Philosophy*

Compass 9, no. 10 (2014): 689–701; William A. Galston, "Realism in Political Theory," *European Journal of Political Theory*, 9, no. 4 (2010): 385–411; Edward Hall, "Bernard Williams and the Basic Legitimation Demand: A Defence," *Political Studies* 63, no. 2 (2015): 466–80; and Janosch Prinz, "Raymond Geuss' Radicalization of Realism in Political Theory," *Philosophy and Social Criticism* 42, no. 8 (2016): 777–96.

2. Williams refers to critical theory's obfuscations in Bernard Williams, *Ethics and the Limits of Philosophy* (Cambridge, MA: Harvard University Press, 1985), 166, 220n; and to its value in Williams, *Truth and Truthfulness: An Essay in Genealogy* (Princeton, NJ: Princeton University Press, 2002), 225–32; and Williams, *In the Beginning Was the Deed: Realism and Moralism in Political Argument*, ed. Geoffrey Hawthorn (Princeton, NJ: Princeton University Press, 2005), 6–7, 14, 82, 89, 93; Guess critiziczes recent critical theory and Rawls in Geuss, *World without Why*, 61, 103; Geuss, "Philosophical Anthropology and Social Criticism." *Reification: A New Look at an Old Idea*, ed. Martin Jay (Oxford: Oxford University Press, 2008), 126; and Geuss, *Reality and Its Dreams* (Cambridge, MA: Harvard University Press, 2016), 81–84.

3. Geuss, *World without Why*, 70.

4. Williams, *Beginning*, 5–6.

5. Hauke Brunkhorst, *Critical Theory of Legal Revolutions: Evolutionary Perspectives* (London: Bloomsbury, 2014), 94.

6. Immanuel Kant *An Answer to the Question: What Is Enlightenment? Practical Philosophy*, trans. Mary J. Gregor. (Cambridge, UK: Cambridge University Press, 1996), 17; see Locke, J. 2015. *An Essay Concerning Human Understanding*, ed. Peter H. Nidditch (Oxford: Oxford University Press, 1979), 43.

7. Geuss, *World without Why*, xvi; and Williams, *Beginning*, 47. See also Brunkhorst, *Critical Theory of Legal Revolutions*, 244.

8. Max Horkheimer and Theodor W. Adorno, *Dialectic of Enlightenment: Philosophical Fragments*, trans. Edmund Jephcott (Stanford, CA: Stanford University Press, 2002), esp. 22, 43, 148ff. For a succinct exegesis, see David Held, *Introduction to Critical Theory: Horkheimer to Habermas* (Berkeley: University of California Press, 1980), 155ff.

9. Geuss, *World without Why*, 104.

10. Charles Taylor, "Socialism and Weltanschauung," in *The Socialist Idea: A Reappraisal*, ed. Leszek Kolakowski and Stuart Hampshire (London: Weidenfeld and Nicolson, 1974), 49; Louis Dumont, *Essays on Individualism: Modern Ideology in Anthropological Perspective* (Chicago: University of Chicago Press, 1986); and Albert O. Hirschman, *The Passions and the Interests: Political Arguments for Capitalism before Its Triumph*, 20th anniv. ed. (Princeton, NJ: Princeton University Press, 1997).

11. See John M. Meyer, *Political Nature: Environmentalism and the Interpretation of Western Thought* (Cambridge, MA: MIT Press, 2001).

12. Benjamin Constant, *Political Writings*, ed. Biancamaria Fontana (Cambridge: Cambridge University Press, 1988), 308–28. Quentin Skinner recognizes Hobbes as the original source of political reflection on such modern dualism. See his *Hobbes and Republican Liberty* (Cambridge: Cambridge University Press, 2008), 126. Quotation from Geuss, *World without Why*, 104.

13. Jürgen Habermas, *The Structural Transformation of the Public Sphere: An Inquiry into a Category of Bourgeois Society*, trans. Thomas Burger and Frederick Lawrence (Cambridge, MA: MIT Press, 1989). See also Richard Sennett, *The Fall of Public Man* (Cambridge: Cambridge University Press, 1977).

14. See Paul James, *Nation Formation: Towards a Theory of Abstract Community* (London: Sage, 1996).

15. Geuss, *World without Why*, 103.

16. Richard Bourke and Raymond Geuss, eds., *Political Judgement: Essays for John Dunn* (Cambridge: Cambridge University Press, 2009), 25.

17. Michel Foucault, *The Birth of Biopolitics*, trans. Graham Burchell (New York: Picador, 2008); Louis Dumont, *From Mandeville to Marx: The Genesis and Triumph of Economic Ideology* (Chicago: University of Chicago Press, 1977); Norberto Bobbio, *The Future of Democracy: A Defence of the Rules of the Game* (Cambridge, UK: Polity, 1991); and Hirschman, *The Passions and the Interests*.

18. To recognize this historical reality is not to endorse it. See Williams, *Beginning*, 5, 82.

19. Ibid., 3ff. (italics and parentheses in original).

20. Ibid., 4ff.

21. Raymond Geuss, *Politics and the Imagination* (Princeton, NJ: Princeton University Press, 2010), 14. See also his essay "Thucydides, Nietzsche, Williams," in *Outside Ethics* (Princeton, NJ: Princeton University Press, 2005), 219ff.

22. Williams, *Beginning*, 78ff.

23. For example, Nancy Fraser, *Unruly Practices: Power, Discourse, and Gender in Contemporary Social Theory* (Cambridge, UK: Polity, 1989).

24. For the distinction between utilitarian individualist and collectivist readings of Hobbes, see Richard Tuck's introduction to Thomas Hobbes, *Leviathan*, ed. Richard Tuck, rev. student ed. (Cambridge: Cambridge University Press, 1996). For discussion of d'Holbach's influence on Enlightenment thought and politics, see, Jacob, *Radical Enlightenment*; and Corey Robin, *Fear: The History of a Political Idea* (Oxford: Oxford University Press, 2004).

25. Raymond Geuss, "Dialectics and the Revolutionary Impulse," in *Cambridge Companion to Critical Theory*, ed. Fred Rush (Cambridge: Cambridge University Press, 2005), 111; Williams, *Beginning*, 43–44; and *Truth and Truthfulness*, 149–71. On Spinoza's central role in the radical Enlightenment and his influence on Rousseau see, Jacob, *Radical Enlightenment*, xi; and Israel, *Radical Enlightenment*, vi. On Rousseau's idea of nature, see Andrew Biro, *Denaturalizing*

Ecological Politics: Alienation from Nature from Rousseau to the Frankfurt School and Beyond (Toronto: University of Toronto Press, 2005), 84ff.

26. On the dialectic of impossibilities and possibilities opened up by the revolutions, see Brunkhorst, *Critical Theory of Legal Revolutions*, 294ff. On Marx's implicit Spinozism, see Louis Althusser, Étienne Balibar, Roger Establet, Jacques Rancière, and Pierre Macherey, *Reading Capital: The Complete Edition*, trans. Ben Brewster and David Fernbach (1965; London: Verso, 2016); and, more recently, Fédéric Lordon, *Willing Slaves of Capital: Spinoza and Marx on Desire* (London: Verso, 2014).

27. Geuss, *World without Why*, 101–2; see also, Taylor, "Socialism and Weltanschauung," 51.

28. Geuss, *World without Why*, 111.

29. Quotation from ibid., 103. See Horkheimer and Adorno, *Dialectic of Enlightenment*; Herbert Marcuse, *Soviet Marxism: A Critical Analysis* (New York: Columbia University Press, 1985); Marcuse, *One Dimensional Man*. (London: Abacus, 1972); Jürgen Habermas, *Toward a Rational Society: Student Protest, Science, and Politics*, trans. Jeremy J. Shapiro (London: Heinemann, 1971); and Habermas *Knowledge and Human Interests*. trans. Jeremy J. Shapiro (Boston: Beacon, 1971).

"Radicalize" in the sense implied by Jacobs and Israel. On the radicalized holism central to Horkheimer and Adorno's iteration of critical theory, see Martin Jay, *Marxism and Totality: The Adventures of a Concept from Lukács to Habermas* (Berkeley: University of California Press, 1984), esp. 223ff. More recently, see also Marcel Stoetzler, "'It Only Needs All': Re-Reading Dialectic of Enlightenment at 70," *Open Democracy*, June 24 2017, 1–16, https://www.opendemocracy.net/can-europe-make-it/marcel-stoetzler/it-only-needs-all-re-reading-dialectic-of-enlightenment-at-70.

30. Claus Offe, "Reflections on the Institutional Self-Transformation of Movement Politics: A Tentative Stage Model," in *Challenging the Political Order: New Social and Political Movements in Democracies*, eds. Russell J. Dalton and Manfred Kuechler (Cambridge: Polity, 1990), 233.

31. Andy Scerri, *Greening Citizenship: Sustainable Development, the State and Ideology* (Basingstoke, UK: Palgrave Macmillan, 2012), 29.

32. On citizenship in the Cold War military-industrial, Keynesian welfare state, see Bryan S. Turner, "Outline of a Theory of Citizenship," in *Citizenship: Critical Concepts*, eds. Bryan S. Turner and Peter Hamilton. (London: Routledge, 1994), 199–226; and Bryan S. Turner, "The Erosion of Citizenship," *British Journal of Sociology* 52, no. 2 (2001): 189–209.

Theodor W. Adorno, Else Frenkel-Brunswik, Daniel J. Levinson, and R. Nevitt Sanford, *The Authoritarian Personality* (New York: Harper, 1950).

33. Max Horkheimer and Theodor W. Adorno, *Dialectic of Enlightenment: Philosophical Fragments*, trans. Edmund Jephcott (Stanford, CA: Stanford University Press, 2002), xi.

34. Williams, *Truth and Truthfulness*, 231. On Williams's point that critique implies a theory of error, and is therefore a negative phenomenon, see Williams, *Truth and Truthfulness*, 232.

Although beyond the ambit of this discussion, there exist a number of interesting parallels between Adorno's philosophical argument and the French school of convention theory. Adorno's tripartite dialectical distinction between nonidentity, identity, and false nonidentity thinking seem reflected in that sociological approach. I am unaware of whether or not convention theorists make this link, but do apply the convention theory terms "critique," "assimilation," and "recuperation" in order to read Adorno's pure philosophy through the lens of Williams's political philosophy. See Luc Boltanski and Laurent Thévenot, *On Justification: Economies of Worth*, trans. Catherine Porter (Princeton, NJ: Princeton University Press, 2006); Boltanski and Eve Chiapello, *The New Spirit of Capitalism*, trans. Gregory Elliott (London: Verso, 2005).

35. Notably, Friedrich Nietzsche's *Beyond Good and Evil* and *The Gay Science* seek to demonstrate the derivation of Enlightenment values from the "eternal" Judeo-Christian distinction between good and evil. As with the Judeo-Christian distinction, the search for meaning as truth implies knowing beforehand and once and for all what is good and what is evil in any situation. Yet "reality" has no reason, right, or good. Reality therefore renders such efforts ironic, such that the pursuit of pure good by the righteous brings terrible consequences. Adorno had witnessed firsthand how the dogged pursuit of essential truth by the self-appointed agents of righteousness led to the marginalizing and, ultimately, excising from humanity of those deemed unfit for good. Hauke Brunkhorst emphasizes Adorno's "admiration" for Nietzsche as an Enlightenment thinker in these terms. See Brunkhorst, *Adorno and Critical Theory* (Cardiff: University of Wales Press, 1999), 53; Friedrich Nietzsche, *Beyond Good and Evil: Prelude to a Philosophy of the Future*, trans. Judith Norman (Cambridge: Cambridge University Press, 2002); Nietzsche, *The Gay Science*, trans. J. Nauckhoff (Cambridge: Cambridge University Press, 2001).

36. Williams, *Truth and Truthfulness*, 258ff.; and Raymond Geuss, *Reality and Its Dreams* (Cambridge, MA: Harvard University Press, 2016), 25ff.

37. Williams, *Beginning*, 1ff (emphasis removed). See also Rossi and Sleat, "Realism in Normative Political Theory," 689–90; Hall, "Williams: A Defence," 466–67.

38. Theodor Adorno, *Negative Dialectics*, trans. E. B. Ashton (London: Routledge & Kegan Paul Ltd., 1973), 8. See also, Marshall Berman, *All That Is Solid Melts Into Air: The Experience of Modernity* (New York: Simon & Schuster, 1982).

39. See Horkheimer's "Atheism and Theism," in *Critique of Instrumental Reason: Lectures and Essays Since the End of World War II* (New York: Seabury, 1974).

40. Theodor W. Adorno, *Minima Moralia: Reflections from Damaged Life*, trans. E. F. N. Jephcott (London: Verso, 1974), 26, 39, 50. See also Brunkhorst,

Adorno and Critical Theory, 36, 114, 130; and Geuss, *World without Why*, 110, citing the German "das bittere Glück des Erkennens," for which the English reference is Adorno, *Minima Moralia*, 26. In German, "Der Distanzierte bleibt so verstrickt wie der Betriebsame; vor diesem hat er nichts voraus als die Einsicht in seine Verstricktheit und das Glück der winzigen Freiheit, die im Erkennen als solchem liegt." Theodor W. Adorno: *Minima Moralia. Reflexionen aus dem beschädigten Leben* (Suhrkamp Verlag, Frankfurt am Main, 1969), 23, 42, 57. See also online a slightly different English translation and the German complete original, available at https://www.marxists.org/reference/archive/adorno/1951/mm/ch01.htm and at http://www.offene-uni.de/archiv/textz/textz_phil/minima_moral.pdf.

41. On Adorno's use of the concept of mimesis as a premodern normative foundation, see Ernesto Verdeja, "Adorno's Mimesis and Its Limitations for Critical Social Thought," *European Journal of Political Theory* 8, no. 4 (2010): 508. On Adorno's view from nowhere, see Brunkhorst, *Adorno and Critical Theory*, 75. Indeed, Simone Chambers argues that for Adorno and the other early critical theorists,

> it was not a political regime that needed overthrowing, so much as the Enlightenment itself. What was called for was a deep, historical transformation akin to the transformation that took place from the Middle Ages to the modern world. Asking critical theorists to propose policies or a political program to bring about this transformation would be like asking Francis Bacon to suggest policies to bring about the Enlightenment. They saw themselves as contributors to a body of knowledge that might (but more likely would not) one day be part of a history of transformation.

Simone Chambers, "The Politics of Critical Theory," in *The Cambridge Companion to Critical Theory*, ed. Fred Rush (Cambridge: Cambridge University Press, 2004), 227.

42. On Adorno's practical actions, see Claus Offe, *Reflections on America: Tocqueville, Weber and Adorno in the United States*, trans. Patrick Camiller (Cambridge, UK: Polity, 2005), 91–92; Brunkhorst, *Adorno and Critical Theory*, 99, 143–44. On Adorno as a democratic theorist, see Shannon Mariotti, *Adorno and Democracy: The American Years* (Lexington: University Press of Kentucky, 2016).

43. On mimetic regression, see Adorno, *Negative Dialectics*, 158, 191.

On the relationship of such regression to Spinozism, and its ultimate expression in the work of Heidegger, see Adorno, *Negative Dialectics*, 21–22, 76.

Adorno's critique of Heidegger's philosophy is fully laid out in Theodor W. Adorno, *The Jargon of Authenticity*, trans. Knut Tarnowski and Frederic Will (London: Routledge and Kegan Paul, 1973).

44. Elsewhere, Adorno argues that progress cannot be understood as a positive category but must rather be understood as a negative category. See

his essay "Progress," in *Can One Live after Auschwitz: A Philosophical Reader*, ed. Rolf Tiedemann (Stanford, CA: Stanford University Press, 2003), 126–45. Colin Koopman regards Williams's interpretation of Nietzschean genealogy as vindicatory; see his *Genealogy as Critique: Foucault and the Problems of Modernity* (Bloomington: Indiana University Press, 2013).

45. Williams, *Beginning*, 1–2; Rossi and Sleat, "Realism in Normative Political Theory," 689; Hall, "Williams: A Defence," 466–67.

46. Horkheimer and Adorno, *Dialectic of Enlightenment*, 3, 29.

47. Adorno, *Negative Dialectics*, 65, 108. Adorno's fear that Heidegger's ontological philosophy operates such that

> The conflict between subjectivity and forms is undiminished, but under the universal rule of forms a consciousness that feels impotent, that has lost confidence in its ability to change the institutions and their mental images, will reverse the conflict into identification with the aggressor (94).

48. Ibid., 15–18, 158–61.

49. Theodor W. Adorno, "Resignation," *Telos*, no. 35 (1978): 165–68; Theodor W. Adorno and Herbert Marcuse, "Correspondence on the German Student Movement," *New Left Review* 233 (1999): 123–36. See also Martin Jay, *The Dialectical Imagination: A History of the Frankfurt School and the Institute of Social Research, 1923–1950* (Berkeley: University of California Press, 1996). More recently, see Stuart Jeffries, *Grand Hotel Abyss: The Lives of the Frankfurt School* (London: Verso, 2016).

50. All quotations in this paragraph are from Offe, "A Tentative Stage Model," 234.

51. Habermas, *Toward a Rational Society*, 121.

52. All quotations are from Geuss, *World without Why*, 111.

53. Joachim Radkau, *Nature and Power: A Global History of the Environment*, trans. Thomas Dunlap (Cambridge: Cambridge University Press, 2008); and Radkau, *The Age of Ecology: A Global History*, trans. Patrick Camiller (Cambridge, UK: Polity, 2014). See also Robert Paehlke, "Eco-History: Two Waves in the Evolution of Environmentalism," *Alternatives* 19, no. 1 (1992): 18–23.

54. Claims that the dawning of a new age, or more specifically a greening of America and establishment of a newly open society that opposed technocracy, abounded in the 1970s. Notable examples include Charles A. Reich, *The Greening of America* (London: Penguin, 1995); Theodore Roszak, *The Making of a Counterculture: Reflections on the Technocratic Society and Its Youthful Opposition*, with new introd. (Berkeley: University of California Press, 1995); and Frank Musgrove, *Ecstasy and Holiness: Counter Culture and the Open Society* (London: Methuen, 1974).

55. Donald Worster, *Nature's Economy: A History of Ecological Ideas* (Cambridge: Cambridge University Press, 1994).

56. Worster is often cited as harbinger of Anthropocene talk. See Andrew Biro, "The Good Life in the Greenhouse? Autonomy, Democracy, and Citizenship in the Anthropocene," *Telos*, no. 172 (2015): 15–37; Matthew Lepori, "There Is No Anthropocene: Climate Change, Species-Talk, and Political Economy," *Telos*, no. 172 (2015): 103–24; Zev Trachtenberg, "The Anthropocene, Ethics, and the Nature of Nature," *Telos*, no. 172 (2015): 38–58; and Timothy W. Luke, "The Anthropocene and Freedom: Terrestrial Time as Political Mystification," *Platypus Review* 60, no. 2 (2013): 2–4.

57. Jedediah Purdy, *After Nature: A Politics for the Anthropocene* (Cambridge, MA: Harvard University Press, 2015), 267, 288.

58. Radkau, *Age of Ecology*, 7–8.

59. Hans Magnus Enzensberger, "A Critique of Political Ecology," *New Left Review*, no. 84 (1974): 3–31.

60. For a summary, see Timothy W. Luke, "Environmentality as Green Governmentality," in *Discourses of the Environment*, ed. Éric Darier (Malden, MA: Blackwell, 1999), 121–51.

61. Timothy W. Luke, "Culture and Politics in the Age of Artificial Negativity," *Telos*, no. 35 (1978): 56–72; "Informationalism and Ecology," *Telos*, no. 56 (1983): 59–73; *Social Theory and Modernity: Critique, Dissent, and Revolution* (Thousand Oaks, CA: Sage, 1990), 12, 159ff.; and *Capitalism, Democracy, and Ecology: Departing from Marx* (Urbana: University of Illinois Press, 1999), 127–31.

62. Radkau, *Age of Ecology*, 427.

63. Quotations are from Radkau, *Age of Ecology*, 9–10, 426.

64. For a similar critique of Radkau's normative defense of incremental localism, see Kate Stevens, "Reviews: An Eco-Contrarian," *New Left Review*, no. 102 (2016): 131–39.

65. Brunkhorst, *Adorno and Critical Theory*, 6. See Ingolfur Blühdorn, "Beyond Criticism and Crisis: On the Post-critical Challenge of Niklas Luhmann," *Debatte* 7, no. 2 (1999): 185–99; Blühdorn, *Post-ecologist Politics: Social Theory and the Abdication of the Ecologist Paradigm* (London: Routledge, 2000); and Ingolfur Blühdorn and Ian Welsh, "Eco-politics Beyond the Paradigm of Sustainability: A Conceptual Framework and Research Agenda," *Environmental Politics* 16, no. 2 (2007): 185–205.

Chapter 2

1. Both deep and social ecologists made much of their links to critical theory. See Steven Vogel, *Against Nature: The Concept of Nature in Critical Theory* (Albany: State University of New York Press, 1996), 6, 8; Brian Tokar,

"On Bookchin's Social Ecology and Its Contributions to Social Movements," *Capitalism, Nature, Socialism* 19, no. 1 (2008): 58; and Robyn Eckersley, *Environmentalism and Political Theory: Toward an Ecocentric Approach* (Albany: State University of New York Press, 1992), 49–50, 98.

2. Daniel T. Rodgers, *The Age of Fracture* (Cambridge, MA: Belknap Press of Harvard University Press, 2012), 3–5; Samuel Moyn, *The Last Utopia: Human Rights in History* (Cambridge, MA: Belknap Press of Harvard University Press, 2010); Kim Phillips-Fein, *Invisible Hands: The Making of the Conservative Movement from the New Deal to Reagan* (New York: Norton, 2009); Steve Fraser, *The Age of Acquiescence: The Life and Death of American Resistance to Organized Wealth and Power* (New York: Little, Brown and Company, 2015); Fraser, *The Limousine Liberal: How an Incendiary Image United the Right and Fractured America* (New York: Basic Books, 2016); Thomas Frank, *One Market under God: Extreme Capitalism, Market Populism, and the End of Economic Democracy* (London: Vintage, 2000); Frank, *The Wrecking Crew: How Conservatives Rule* (New York: Henry Holt, 2008); Jane Mayer, *Dark Money: The Hidden History of the Billionaires behind the Rise of the Radical Right* (New York: Doubleday, 2016); and Nancy MacLean, *Democracy in Chains: The Deep History of the Radical Right's Stealth Plan for America* (New York: Viking, 2017).

3. Barbara Epstein, *Political Protest and Cultural Revolution: Nonviolent Direct Action in the 1970s and 1980s* (Berkeley: University of California Press, 1991); Robert Gottlieb, *Forcing the Spring: The Transformation of the American Environmental Movement*, 2nd ed. (Washington, DC: Island Press, 2005). See also Robert J. Brulle, *Agency, Democracy, and Nature: The U.S. Environmental Movement from a Critical Theory Perspective* (Cambridge, MA: MIT Press, 2000); Charles T. Rubin, *The Green Crusade: Rethinking the Roots of Environmentalism* (Lanham, MD: Rowman & Littlefield, 1993); Kirkpatrick Sale, *The Green Revolution: The American Environmental Movement, 1962–1992* (New York: Hill and Wang, 1994); and Philip Shabecoff, *A Fierce Green Fire: The American Environmental Movement* (New York: Hill and Wang, 1994).

4. Gottlieb, *Forcing the Spring*, 139.

5. Quotes from, ibid., 149–50. President Nixon's 1970 *State of the Union Address* is quoted in ibid., 152. For an excerpt from the address, go to https://www.youtube.com/watch?v=82Fl4vhbUys.

On the Nixon administration's public attitude to the environmental problem, see James Morton Turner, "'The Specter of Environmentalism': Wilderness, Environmental Politics, and the Evolution of the New Right," *Journal of American History* 96, no. 1 (2009): 129.

6. On broadening the definition of domination, Edward P. Morgan, *The Sixties Experience: Hard Lessons about Modern America* (Philadelphia: Temple University Press, 1991), 12. See also Bruce J. Schulman, *The Seventies: The Great Shift in American Culture, Society, and Politics* (New York: Da Capo, 2002),

90–91; Brulle, *Agency, Democracy, and Nature*, 168ff.; and Timothy W. Luke, "Informationalism and Ecology." *Telos*, no. 56 (1983): 59–73. For a contemporaneous account, see Douglas W. Scott, "Student Activism on Environmental Crisis," *Living Wilderness* 34 (1970): 231–42.

8. Murray Bookchin, *The Ecology of Freedom: The Emergence and Dissolution of Hierarchy* (Oakland, CA: AK Press, 2005), 109, 413–14.

9. Arne Naess, *Ecology, Community, and Lifestyle: Outline of an Ecosophy*, trans. David Rothenberg (Cambridge: Cambridge University Press, 1993), 204.

10. Murray Bookchin, "Beyond Neo-Marxism," *Telos* 1978, no. 36 (1978): 19.

11. Bill Devall and George Sessions, *Deep Ecology: Living as if Nature Mattered* (Salt Lake City, UT: Peregrine Smith, 1985), 65–66, 100.

12. Murray Bookchin and Dave Foreman, *Defending the Earth: A Debate*, ed. Steve Chase (Montreal: Black Rose Books, 1991), 81.

13. See Wouter J. Hanegraaff, *New Age Religion and Western Culture: Esotericism in the Mirror of Secular Thought* (New York: State University of New York Press, 1998); Paul Heelas, *The New Age Movement: The Celebration of the Self and the Sacralization of Modernity* (Oxford: Blackwell, 1996); Musgrove, *Ecstasy and Holiness*.
See also Arran Gare's work on the philosophical origins of the green-left synthesis, *Postmodernism and the Environmental Crisis* (London: Routledge, 1995); "From Kant to Schelling to Process Metaphysics: On the Way to Ecological Civilization," *Cosmos and History* 7, no. 2 (2011): 26–69; "The Grand Narrative of the Age of Re-embodiments: Beyond Modernism and Postmodernism," *Cosmos and History* 9, no. 1 (2013), 327–57; "Deep Ecology, the Radical Enlightenment and Ecological Civilization," *Trumpeter* 30, no. 2 (2014), 184–205; and *The Philosophical Foundations of Ecological Civiliation: A Manifesto for the Future* (London: Routledge, 2016).

14. For a detailed account of deep ecological Spinozism, see Bron Taylor, "Earth and Nature-Based Spirituality (Part I): From Deep Ecology to Radical Environmentalism," *Religion* 31, no. 2 (2001): 175–93. The social ecologist Murray Bookchin forthrightly rejected the deep ecologists' *interpretation* of Spinoza—"a bizarre mix of Buddhism, Taoism, Native American beliefs, Heidegger, and Spinoza . . ."—but did so on grounds that they had misinterpreted the philosopher's argument. Indeed, Bookchin regarded an accurate reading of Spinoza as in fact "bearing a stronger affinity with [the social ecological] commitment to organic entelechies and dialectical reason." See Bookchin, "Recovering Evolution: A Reply to Eckersley and Fox," *Environmental Ethics* 12, no. 3 (1990): 253–74; and "Will Ecology Become 'the Dismal Science,'" *Progressive*, December 1991, 18–21.

15. For a comprehensive list of influences on the deep ecologists, see Timothy W. Luke, "'Deep Ecology: Living as if Nature Mattered': Devall and Sessions on Defending the Earth," *Organization and Environment* 15, no. 2 (2002):

182. See also James Lovelock and Sidney Epton, "The Quest for Gaia," *New Scientist* (February 6, 1975): 304–6; Gregory Bateson, *Steps to an Ecology of Mind* (Frogmore, UK: Paladin, 1973); J. Baird Callicott, "Intrinsic Values, Quantum Theory, and Environmental Ethics," *American Philosophical Quarterly* 21 (1985): 299–309; and Callicott, "The Metaphysical Implications of Ecology," *Environmental Ethics* 8, no. 4 (1986): 301–16. Quotes on the anthropology of social ecology in Murray Bookchin, *Post-scarcity Anarchism* (Montreal: Black Rose Books, 1986), 58. See also Bookchin, "Beyond Neo-Marxism."

16. Aldo Leopold, *A Sand County Almanac, with Essays on Conservation from Round River* (New York: Sierra Club, 1966), 187; Devall and Sessions, *Deep Ecology*, 65; Arne Naess, "The Shallow and the Deep, Long-Range Ecology Movement: A Summary," *Inquiry* 16 (1973): 100. For an overview, see Peter Hay, *Main Currents in Westerm Environmental Thought* (Sydney: University of New South Wales Press, 2002); Gary Snyder, *The Practice of the Wild* (San Francisco: Northpoint, 1990); Bookchin, *Post-scarcity Anarchism*, 65.

17. As Epstein shows, the idea of prefiguring a desired social order in the here and now had been taken up by the original New Left of the early 1960s, notably by the Students for a Democratic Society and Quaker civil rights activists: *Political Protest and Cultural Revolution*, 16–17, 261–62, 76–78. See also Wini Breines, *Community and Organization in the New Left, 1962–1968: The Great Refusal*, 2nd ed. (New Brunswick, NJ: Rutgers University Press, 1989), 6. On Gramscian and autonomous Marxist debates over prefiguration as a specific response to the false dualisms of capitalism: see Carl Boggs, "Marxism, Prefigurative Communism, and the Problem of Workers' Control," *Radical America* 11, no. 6 (1977), 113.

18. Andrew Light and Holmes Rolston III, "Introduction: Ethics and Environmental Ethics," in *Environmental Ethics: An Anthology*, ed. Andrew Light and Holmes Rolston III (Maldon, MA: Blackwell, 2003), 1–11; and Leopold, *A Sand County Almanac*, 204. See also J. Baird Callicott, *In Defense of the Land Ethic: Essays in Environmental Philosophy* (Albany: State University of New York Press, 1989); Callicott, *Beyond the Land Ethic: More Essays in Environmental Philosophy* (Albany: State University of New York Press, 1999); Roderick Nash, "Aldo Leopold's Intellectual Heritage," in *Companion to "A Sand County Almanac": Interpretive and Critical Essays*, ed. J. Baird Callicott (Madison: University of Wisconsin Press, 1987), 63–88; and Bookchin, *Ecology of Freedom*, 97.

19. See, Ulrich Beck, *Risk Society: Towards a New Modernity*, trans. Mark Ritter (London: Sage, 1992).

20. Devall and Sessions, *Deep Ecology*, 65–66; also cited in Luke, "Devall and Sessions on Defending the Earth."

21. Bookchin, *Post-Scarcity Anarchism*, 67.

22. Bookchin and Foreman, *Defending the Earth*, 78, 80–81.

23. Bookchin, *Post-scarcity Anarchism*, 67, 101–2.

24. Ibid., 67, 187.

25. Marcuse, *An Essay on Liberation* (London: Penguin, 1969); Dumont, *Essays on Individualism*, 217. See also David Harvey, "The Nature of Environment: The Dialectics of Social and Environmental Change," *Socialist Register* 29 (1993): 1–51; Mike Davis, "The Political Economy of Late-Imperial America," *New Left Review*, no. 143 (1984): 6–38; Thomas Frank, *The Conquest of Cool: Business Culture, Counterculture, and the Rise of Hip Consumerism* (Chicago: University of Chicago Press, 1997); and Boris Frankel, *The Post-industrial Utopians* (Cambridge, UK: Polity, 1987).

26. Charles Taylor, *Philosophical Arguments* (Cambridge, MA: Harvard University Press, 1995), 100; Reich, *The Greening of America*; Roszak, *The Making of a Counterculture*. See Ellen Meiksins Wood, "A Chronology of the New Left and Its Successors, or: Who's Old-Fashioned Now?," *Socialist Register* 31 (1995), 22–49; Ira Katznelson, "From the Streets to the Lecture Hall: The 1960s," *Daedelus* 126, no. 1 (1997): 311–32; Taylor, "Socialism and Weltanschauung," 48; Taylor, *Philosophical Arguments*, 100ff.; Taylor, *Sources of the Self: The Making of the Modern Identity* (Cambridge, MA: Harvard University Press, 1989), 102, 342–43. On the use of MacIntyre's philosophy, see Arran Gare "Macintyre, Narratives, and Environmental Ethics," *Environmental Ethics* 20, no. 1 (1998): 3–21; and Mick Smith, *An Ethics of Place: Radical Ecology, Postmodernity, and Social Theory* (Albany: State University of New York Press, 2001).

27. It is necessary here to distinguish clearly between the rights-based contractualism advocated by the communitarian's interlocutors, such as liberal pluralist John Rawls and libertarian Robert Nozick, and the collective contractualism that I discuss in chapter 1. See John Rawls, *A Theory of Justice*, rev. ed. (Cambridge, MA: Belknap Press of Harvard University Press, 2005); and Robert Nozick, *Anarchy, State, and Utopia* (New York: Basic Books, 1974).

28. Taylor, *Philosophical Arguments*, 7.

29. Charles Taylor, *Hegel* (Cambridge: Cambridge University Press, 1975). Geuss makes a similar connection in, *World without Why*, 99–100.

30. Alongside Judi Bari, perhaps the most well known was John Zerzan, *Elements of Refusal* (St Louis: Left Bank Books, 1988).

31. Taylor, "Socialism and Weltanschauung," 49–50.

32. Ibid., 50.

33. Brunkhorst, *Adorno and Critical Theory*, 75; Verdeja, "Adorno's Mimesis and Its Limitations for Critical Social Thought," 508.

34. Geuss, *World without Why*, 104.

35. All quotes from Taylor, *Philosophical Arguments*, 181, 182, 202.

36. Taylor, *Sources of the Self*, 8.

37. Raymond Geuss, *Philosophy and Real Politics* (Princeton, NJ: Princeton University Press, 2008), 9ff. Rawls, Nozick, and indeed Habermas are frequent targets for Geuss. See also his *Outside Ethics*; *Reality and Its Dreams*; and *History*

and Illusion in Politics (Cambridge: Cambridge University Press, 2001). Bernard Williams's reviews of work by Taylor, MacIntyre, Rawls, and Nozick are reproduced in *Essays and Reviews, 1959–2002* (Princeton, NJ: Princeton University Press, 2014) See also *Beginning*, 13ff., 29ff., 130ff. For a defense of Ralws, see Alan Thomas, "Rawls and Political Realism: Realistic Utopianism or Judgement in Bad Faith?" *European Journal of Political Theory* 16, no. 3 (2017): 304–24.

38. On the tripartite distinction, see Alasdair MacIntyre, *Whose Justice? Which Rationality?* (London: Duckworth, 1988), 1–4; and *After Virtue*, 2nd ed. (Notre Dame, IN: University of Notre Dame Press, 1986), 3, 6–7. See also Fred Block and Margaret R. Somers, *The Power of Market Fundamentalism: Karl Polanyi's Critique* (Cambridge, MA: Harvard University Press, 2014), 38.

39. See Luke, *Capitalism, Democracy, and Ecology*; and "Environmentality as Green Governmentality."

40. Williams, *Beginning*, 1–2.

41. Ibid., 24, 33; Williams, *Ethics and the Limits of Philosophy*, 168, 197; and Geuss, *Philosophy and Real Politics*, 60ff.

42. Radkau, *Age of Ecology*, 427.

43. Gottlieb, *Forcing the Spring*, 141ff.

44. William Ophuls, *Ecology and the Politics of Scarcity: Prologue to a Political Theory of the Steady State* (San Francisco: W. H. Freeman, 1977); Paul R. Ehrlich, *The Population Bomb* (New York: Ballantine, 1968); and Garrett Hardin, "The Tragedy of the Commons," *Science* December 13, 1968, 1243–48.

45. Mike Mills, "Green Democracy: The Search for an Ethical Solution," in *Democracy and Green Political Thought: Sustainability, Rights, and Citizenship*, ed. Brian Doherty and Marius de Geus (London: Routledge, 1996), 98 (Ophuls quoted by Mills). For Hardin's opposition to utilitarianism, see the conclusion to his "The Tragedy of the Commons."

46. Williams, *Beginning*, 3.

47. Ibid.; also, Enzo Rossi, "Justice, Legitimacy and (Normative) Authority for Political Realists," *Critical Review of International Social and Political Philosophy* 15, no. 2 (2012): 149–64.

48. See note 2, this chapter.

49. This was the task set for Greens and the new social movements generally by Barry Commoner in his widely read *Closing the Circle: Nature, Man, and Technology* (New York: Knopf, 1971).

Chapter 3

1. See James O'Connor, *The Fiscal Crisis of the State* (New York: Transaction, 2009); Raymond Vernon, ed. *The Oil Crisis* (New York: Norton, 1976); and William J. Baumol, "Macroeconomics of Unbalanced Growth: The Anatomy

of Urban Crisis," *American Economic Review* 57, no. 3 (1968): 415–26; Michael Crozier, Samuel P. Huntington, and Joji Watanuki, *The Crisis of Democracy: Report on the Governability of Democracies to the Trilateral Commission* (New York: New York University Press, 1975); and Habermas, *Legitimation Crisis*, 20–24; Claus Offe, *Contradictions of the Welfare State* (Cambridge, MA: MIT Press, 1984).

2. John H. Schaar, "Legitimacy and the Modern State," in *Power and Community: Dissenting Essays in Political Science*, ed. Philip Green and Sanford Levinson (New York: Vintage, 1970), 317.

3. For a contemporaneous view, see Charles Denby, "Workers Battle Automation," in *The New Left: A Collection of Essays*, ed. Priscilla Long (Boston: Porter Sargent, 1969). For a look back by someone involved in such efforts, see the inroduction to Gar Alperovitz, *America beyond Capitalism: Reclaiming Our Wealth, Our Liberty, and Our Democracy*, 2nd ed. (Cleveland: Democracy Collaborative, 2011).

4. See Ronald Inglehart, *The Silent Revolution: Changing Values and Political Styles among Western Publics* (Princeton, NJ: Princeton University Press, 1977); Herbert Marcuse, *Eros and Civilization: A Philosophical Inquiry into Freud* (London: Sphere, 1970); Marcuse, *One Dimensional Man*; Reich, *The Greening of America*; Roszak, *The Making of a Counterculture*; Hanegraaf, *New Age Religion and Western Culture*; Heelas, *The New Age Movement*; Musgrove, *Ecstasy and Holiness*.

5. See David Harvey, *The Condition of Postmodernity: An Enquiry into the Origins of Cultural Change* (Oxford: Blackwell, 1990); and Robert Brenner, *The Boom and the Bubble: The US in the World Economy* (London: Verso, 2005).

6. Jürgen Habermas, "New Social Movements," *Telos*, no. 49 (1981): 33. See also *Lifeworld and System: A Critique of Functionalist Reason*, vol. 2 of *The Theory of Communicative Action*, trans. Thomas McCarthy (Boston: Beacon, 1987), 392–96.

7. Habermas, *Toward a Rational Society*, 121–22. One contribution to the more recent political history of the 1970s that makes very similar observations is Schulman, *The Seventies*.

8. Nancy Fraser, "Legitimation Crisis? On the Political Contradictions of Financialized Capitalism," *Critical Historical Studies* 2, no. 2 (2015): 165, 173.

9. Held, *Introduction To Critical Theory*, 250–51. See also Theodor W. Adorno and Herbert Marcuse, "Correspondence on the German Student Movement," *New Left Review* 233 (1999): 123–36.

10. Habermas, *Legitimation Crisis*, 123–24.

11. Ibid. Habermas' contends that the simultaneous crisis of economic growth (50–61), the administrative crisis of rationality (61–68), and most importantly the legitimation crisis of the state (68–75) could be traced to the motivational crisis (75–92) that Horkheimer and Adorno had theorized as the "death of the bourgeois individual" in their *Dialectic of Enlightenment* (125–26).

12. See also, Anthony Giddens, *The Class Structure of the Advanced Societies* (London: Hutchinson & Co. Ltd, 1974), 275–94.

13. Frances Fox Piven, *Challenging Authority: How Ordinary People Change America* (Lanham, MD: Rowman and Littlefield, 2006); Phillips-Fein, *Invisible Hands*; Phillips-Fein, "Business Conservatives and the Mont Pèlerin Society," in *The Road from Mont Pèlerin*, ed. Philip Mirowski and Dieter Plehwe (Cambridge, MA: Harvard University Press, 2009), 280–301; Rob Van Horn, and Philip Mirowski, "The Rise of the Chicago School of Economics and the Birth of Neoliberalism," in *The Road from Mont Pèlerin*, ed. Philip Mirowski and Dieter Plehwe, 138–78; Mayer, *Dark Money*; Hacker and Pierson, *American Amnesia*; Lafer, *The One Percent Solution*; also, Pierre Dardot and Christian Laval, *The New Way of the World: On Neoliberal Society*, trans. Gregory Elliott (London: Verso, 2014); and Justin Vaïsse, *Neoconservatism: The Biography of a Movement*, trans. Arthur Goldhammer (Cambridge, MA: Belknkap Press of Harvard University Press, 2010).

14. Habermas, *Legitimation Crisis*, 41, 73.

15. Ibid., 68–70.

16. For Habermas's critique of Horkheimer, Adorno, Marcuse, and those he labels the new conservatives of postmodernism, see *The Philosophical Discourse of Modernity: Twelve Lectures*, trans. Frederick G. Lawrence (Cambridge, MA: MIT Press, 1990).

17. On Habermas's fully developed theory, see Jürgen Habermas, *Knowledge and Human Interests*, trans. Jeremy J. Shapiro (Boston: Beacon, 1971); *Reason and the Rationalization of Society*, vol. 1 of *The Theory of Communicative Action*, trans. Thomas McCarthy (Boston: Beacon, 1984); *Lifeworld and System*; and *Between Facts and Norms: Contributions to a Discourse Theory of Law and Democracy*, trans. William Rehg (Cambridge, MA: MIT Press, 1994).

Habermas's approach would be elaborated theoretically and as praxis by numerous advocates of "deliberative democratic theory"; for example, John S. Dryzek, *Deliberative Democracy and Beyond: Liberals, Critics, Contestations* (Oxford: Oxford University Press, 2000); Dryzek, *Deliberative Global Politics: Discourse and Democracy in a Divided World* (Cambridge, UK: Polity, 2006); Dryzek, "Democratization as Deliberative Capacity Building," *Comparative Political Studies* 42, no. 11 (2009): 1379–402; Dryzek, *Foundations and Frontiers of Deliberative Governance* (Oxford: Oxford University Press, 2010); Robyn Eckersley, "Deliberative Democracy, Ecological Representation and Risk: Towards a Democracy of the Affected," in *Democratic Innovation: Deliberation, Representation and Association*, ed. Michael Saward (London: Routledge, 2000), 117–32; Robert E. Goodin, *Innovating Democracy: Democratic Theory and Practice after the Deliberative Turn* (Oxford: Oxford University Press, 2008); Robert E. Goodin and John S. Dryzek, "Deliberative Impacts: The Macro-political Uptake of Mini-publics," *Politics and Society* 34, no. 2 (2006): 219–44; Carolyn M. Hendriks, "Institutions of Deliberative Democratic Process and Interest Groups: Roles, Tensions and Incentives," *Australian Journal of Public Administration* 61, no. 1 (2002): 64–75; Hendriks, "When the Forum Meets Interest Politics: Strategic Uses of Public Deliberation,"

Politics and Society 34, no. 4 (2006): 571–602; Hendriks, "Deliberative Governance in the Context of Power," *Policy and Society* 28, no. 3 (2009): 173–84; Carolyn M. Hendriks, John S. Dryzek, and Christian Hunold, "Turning up the Heat: Partisanship in Deliberative Innovation," *Political Studies* 55, no. 2 (2007): 632–83; and Judith E. Innes and David E. Booher, "Collaborative Policymaking: Governance through Dialogue," in *Deliberative Policy Analysis: Understanding Governance in the Network Society*, ed. Maarten Hajer and Hendrik Wagenaar (Cambridge: Cambridge University Press, 2003), 33–59.

18. Steven Vogel, *Against Nature: The Concept of Nature in Critical Theory* (Albany: State University of New York Press, 1996), 109–10. See also C. Fred Alford, *Science and the Revenge of Nature: Marcuse and Habermas* (Gainesville, FL: University Presses of Florida, 1985).

19. Most clearly articulated in Habermas, *Between Facts and Norms*.

20. On the destruction of the environment or nature, see Habermas, *Legitimation Crisis*, 41–46; *Lifeworld and System*, 394. On the new social movements as self-misunderstood, see *Lifeworld and System*, 395–96. See also Habermas's direct riposte to Marcuse's enthusiasm for the new social movements and broader counterculture, *Toward a Rational Society*, 86–90, 121–23, which echoed Adorno's similar ambivalence and disagreement with Marcuse; see Adorno and Marcuse, "Correspondence on the German Student Movement."

21. Habermas derided "green Romantics" for eschewing efforts to render politics ethical in favor of efforts to popularize ecocentric "moral outrage": Habermas, "A Reply to My Critics," in *Habermas: Critical Debates*, ed. John B. Thompson and David Held (London: Macmillan, 1982), 219–83. See Vogel, *Against Nature*, 153.

22. Habermas, "Reply," 244–45; Vogel, *Against Nature*, 140 (emphasis in original).

23. Williams, *Truth and Truthfulness*, 225–26, 229; Raymond Geuss, *The Idea of a Critical Theory: Habermas and the Frankfurt School* (Cambridge: Cambridge University Press, 1981), esp. 64ff.; Geuss, *Morality, Culture, and History: Essays on German Philosophy* (Cambridge: Cambridge University Press, 1999), 59–61; Geuss, *Politics and the Imagination*, 77ff.; Geuss, *Outside Ethics*, 92; Geuss, *History and Illusion in Politics*, 79.

24. See note 34, chapter 1.

25. Michael Crozier, Samuel P. Huntington, and Joji Watanuki, *The Crisis of Democracy: Report on the Governability of Democracies to the Trilateral Commission* (New York: New York University Press, 1975).

26. See Kim Phillips-Fein, "Conservatism: A State of the Field," *Journal of American History* 98, no. 3 (2011); also Piven, *Challenging Authority*, 99, 130; and Phillips-Fein, "Business Conservatives," 283.

27. David Vogel, *Fluctuating Fortunes: The Political Power of Business in America* (New York: Basic Books, 1989), 59, cited in Jacob S. Hacker and Paul

Pierson, *Winner-Take-All Politics: How Washington Made the Rich Richer—and Turned Its Back on the Middle Class* (New York: Simon and Schuster, 2010), 176.

28. Piven, *Challenging Authority*, 131. The depth of concern expressed by business groups across the United States and other Western advanced industrial societies was well-documented at the time and more recently. See, for example, Mike Davis, "The New Right's Road to Power," *New Left Review*, no. 128 (1981): 43; Boltanski and Chiapello, *The New Spirit of Capitalism*, 191; Dardot and Laval, *New Way of the World*, 151; William K. Carroll and Jean Philippe Sapinski, "The Global Corporate Elite and the Transnational Policy-Planning Network, 1996–2006: A Structural Analysis," *International Sociology* 25, no. 4 (2010): 502; Stephen Gill, *American Hegemony and the Trilateral Commission* (Cambridge: Cambridge University Press, 1990); Mirowski and Plehwe, *The Road from Mont Pèlerin*, 13; Hacker and Pierson, *Winner-Take-All Politics*, 116ff.; Mayer, *Dark Money*, 72–73; Ward Churchill and Jim Vander Wall, *The Cointelpro Papers: Documents from the FBI's Secret Wars against Dissent in the United States* (Boston: South End Press, 1990); Holly Sklar, *Reagan, Trilateralism, and the Neoliberals* (Boston: South End Press, 1986); Leslie Sklair, *The Transnational Capitalist Class* (Oxford: Blackwell, 2001); MacLean, *Democracy in Chains*, 115ff.; S. M. Amadae, *Rationalizing Capitalist Democracy: The Cold War Origins of Rational Choice Liberalism* (Chicago: University of Chicago Press, 2003); and Amadae, *Prisoners of Reason: Game Theory and Neoliberal Political Economy* (New York: Cambridge University Press, 2016); and Gordon Lafer, *The One Percent Solution: How Corporations Are Remaking America One State at a Time* (Ithaca, NY, Cornell University Press/ILR Press, 2017).

29. MacLean, *Democracy in Chains*, 116.

30. Lewis F. Powell, "Confidential Memorandum: Attack on American Free Enterprise System," 4.

31. Crozier, Huntington, and Watanuki, *Crisis of Democracy*, 113. Incidentally, Ronald Inglehart's problematic work on 'postmaterial values' is referred to directly in the *Crisis of Democracy* report, 6–7. See also Sklar, *Reagan, Trilateralism and the Neoliberals*; and Gill, *American Hegemony and the Trilateral Commission*. To this extent, the rise of the New Right was the product of conversations amongst conservative parties across the West at the time, each seeking a response to the crisis.

32. Compare Powell, "Confidential Memorandum: Attack on American Free Enterprise System," 5, 10 with Phillips-Fein, *Invisible Hands*, 158. See also Vaïsse, *Neoconservatism*; and Adam Hilton, "Searching for a New Politics: The New Politics Movement and the Struggle to Democratize the Democratic Party, 1968–1978," *New Political Science* 38, no. 2 (2016): 141–59; Bruce Miroff, *The Liberals' Moment: The McGovern Insurgency and the Identity Crisis of the Democratic Party* (Lawrence: University Press of Kansas, 2007); Mark Kurlansky, *1968: The Year That Rocked the World* (London: Vintage, 2004), 360–61; Frank, *One*

Market under God, 26–27; and Frances Fox Piven, and Richard A. Cloward, *Poor People's Movements: Why They Succeed, How They Fail* (New York: Vintage, 1979), 319–20.

33. On the establishment of the pollution control apparatus, see Gottlieb, *Forcing the Spring*, 152, 157. Interestingly, the Powell memo refers directly to consumer rights and anti-pollution activist Ralph Nader: Powell, "Confidential," 6. On the broader threat posed by green-left mordnierst faction within the by then broadly radical and reformist Green movement, see John S. Dryzek, et al., *Green States and Social Movements: Environmentalism in the United States, United Kingdom, Germany, and Norway* (Oxford: Oxford University Press, 2003), 59.

34. Nixon's quotation in full, "The great question of the 70s is, 'Shall we surrender to our surroundings or shall we make our peace with nature and begin to make reparations for the damage we have done to our air, to our land, and to our water.'" See, https://www.youtube.com/watch?v=82Fl4vhbUys.

On the Nixon Administration's preemption of the environmetnal issue, see *Forcing the Spring*, 152.

35. Quotations in this paragraph are from Rosenbaum, *Environmental Politics and Policy*, 10, 11; and Gottlieb, *Forcing the Spring*, 170. Over a period of seven years, the Nixon, Ford, and Carter administrations established the federal environmental policy regime, beginning with the advisory Council on Environmental Quality and National Environmental Policy Act (1969), the Environmental Protection Authority (EPA), the Clean Air at (1970), the Clean Water Act (1972), the Endangered Species Act and the Coastal Zone Management Act (1973), the Safe Drinking Water Act (1974), the Toxic Substances Control Act and Resource Conservation and Recovery Act (1976), the Surface Mining Control and Reclamation Act (1977), various amendments to these, and the establishment of the Comprehensive Environmental Response, Compensation, and Liability Act (the "Superfund" Act, 1977). On the implications of Nixon's strategy of driving a wedge between moderate and radical Greens, see Christian Hunold and John S. Dryzek, "Green Political Theory and the State: Context Is Everything," *Global Environmental Politics* 2, no. 3 (2002): 17–39; Radkau, *Age of Ecology*, 304, 306; and Dobson, *Green Political Thought*, 46, 129–30.

36. Quotations in this paragraph are from Rosenbaum, *Environmental Politics and Policy*, 11, 168. See also Burton I. Kaufman, *The Presidency of James Earl Carter, Jr.* (Lawrence: University Press of Kansas, 1993); and William E. Leuchtenburg, "Jimmy Carter and the Post–New Deal Presidency," in *The Carter Presidency: Policy Choices in the Post–New Deal Era*, ed. Gary M. Fink and Hugh Davis Graham (Lawrence: University Press of Kansas, 1998).

37. For Holly Sklar, "Jimmy Carter was the first neoliberal President. He made fiscal conservatism the bipartisan alternative to the Welfare State"; Sklar, *Reagan, Trilateralism and the Neoliberals*, 6. On the mobilization of citizens who had always rejected radicalism and reformist liberalism, see Phillips-Fein, "Con-

servatism," 724. See also Steve Fraser and Gary Gerstle, eds., *The Rise and Fall of the New Deal Order, 1930–1980* (Princeton, NJ: Princeton University Press, 1989); Janice Stein, *Pivotal Decade: How the United States Traded Factories for Finance in the Seventies* (New Haven, CT: Yale University Press, 2010); Jefferson Cowie, *Stayin' Alive: The 1970s and the Last Days of the Working Class* (New York: New Press, 2010).

38. For a sympathetic and contemporaneous account, see Paul Gottfried and Thomas Fleming, *The Conservative Movement* (Boston: Twayne, 1988).

39. See Davis, "Late-Imperial America"; and Harvey, *The Condition of Postmodernity*.

40. See Andy Scerri, "Reaganomics," in *Wiley Encyclopedia of Social Theory* vol. 4, ed. Brian S. Turner (New York: Wiley, 2017), 1939–41.

41. See transcript of Minnesota Republican congress member Vin Weber interviewed by P. J. Boyer for "The Long March of Newt Gingrich," *Frontline*, PBS, January 16, 1996, https://www.pbs.org/wgbh/pages/frontline/newt/newtintwshtml/weber.html.

42. Reagan quoted in Turner, "The Specter of Environmentalism," 134.

43. No doubt in part due to the mainstream lobby's refusal to disavow the radical green-left activism such as the ecotage practiced by Earth First! See Radkau, *Age of Ecology*, 306. On the Wise Use dependence on industry funds, see Turner, "The Specter of Environmentalism," 139. For a book length account of the links between the well-funded New Right and Wise Use and similar faux-grassroots or astroturf movements, see David Helvarg, *The War against the Greens: The "Wise-Use" Movement, the New Right, and the Browning of America* (Boulder, CO: Johnson Books, 2004).

44. Turner, "The Specter of Environmentalism," 139, 41.

45. Gottlieb, *Forcing the Spring*, 167, 218.

46. Turner, "The Specter of Environmentalism," 141–42.

47. The New Right would hone and roll out just such a strategy over the next decades, at the national scale, for example, in relation to welfare and social security reform, and at the scale of state and city government, for example, in relation to public transit. In relation to institutions such as Medicare and Social Security, the New Right would self-present as seeking to reform them while holding to the ultimate aim of destroying these widely popular institutions. See MacLean, *Democracy in Chains*, 194.

48. During a televised debate in the lead up to the 1980 election. See "Ronald Reagan: There You Go Again," YouTube video, last modified January 4, 2008, https://www.youtube.com/watch?v=Wi9y5-Vo61w.

49. On Hayek as guru, see Samuel Moyn, "A Powerless Companion: Human Rights in the Age of Neoliberalism," *Law and Contemporary Problems* 77, no. 4 (2014): 148. On his leadership of the neoliberal thought collective and assistance from the business lobby, see Van Horn and Mirowski, "The Rise

of the Chicago School," 138. On business lobby support for the New right in general and the effects of this support on the movement's public presence, see Frank, *One Market under God*, 114; Rodgers, *The Age of Fracture*, 29; and Block and Somers, *The Power of Market Fundamentalism*, 152.

50. On Hayek's antirationalism and its echoes in business circles, see Van Horn and Mirowski, "The Rise of the Chicago School," 140 and Phillips-Fein, "Conservatism," 735; For a sympathetic account of Hayek's skepticism in relation to Enlightenment values and rationalism in particular, see Bruce Caldwell, introduction to *The Road to Serfdom: Text and Documents*, ed. Bruce Caldwell, definitive ed. (Chicago: University of Chicago Press, 2007). On the road to serfdom, spontaneous order of the cosmos and mutual benefits, Friedrich Hayek, *The Road to Serfdom* (London: Routledge & Kegan Paul, 1979); quotes from *Studies in Philosophy, Politics and Economics* (London: Routledge and Kegan Paul, 1967), 163; and *The Fatal Conceit: The Errors of Socialism*, ed. William Warren Bartley (London: Routledge, 1988), 25.

51. Andy Denis, "A Century of Methodological Individualism, Part 2: Mises and Hayek," (Deptartment of Economics, City University London, 2010), 5.

52. Friedrich Hayek, *The Sensory Order: An Inquiry into the Foundations of Theoretical Psychology* (London: Routledge and Kegan Paul, 1952), 47.

53. Hayek, *Studies*, 70.

54. On Hayek's holism, Andy Denis, "Two Rhetorical Strategies of *Laissez-Faire*," *Journal of Economic Methodology* 11, no. 3 (2004): 348. On Hayek's philosophy as a reflection and sophistication of the business point of view, Phillips-Fein, "Business Conservatives," 282. On Hayek's critique of modern legislative artificialism and the gradualism of the rule of law, Hayek, *Studies*, 113 (paraphrased in Dardot and Laval, *New Way of the World*, 127, 128, 129); for a similar analysis of Schmitt and Lenin, see William E. Scheuerman, W. *Between the Norm and the Exception: The Frankfurt School and the Rule of Law* (Cambridge, MA: MIT Press, 1994), 38.

55. Irving Kristol, "Ideology and Supply-Side Economics," *Commentary* 71, no. 4 (1981): 48.

56. Hayek, *Fatal Conceit*, 20.

57. Hayek, *Studies*, 85.

58. The dissemination and bowdlerization process is explained in Elisabeth K. Chaves, *Reviewing Political Criticism: Journals, Intellectuals, and the State* (Burlington, VT: Ashgate, 2015), 75–76; Kristol, "Ideology and Supply-Side Economics," 48.

59. Dardot and Laval, *New Way of the World*, 130.

60. Denis, "Rhetorical Strategies," 349 (italics in original).

61. Quotations in this paragraph are from Andy Denis, "Was Hayek a Panglossian Evolutionary Theorist: A Reply to Whitman," *Constitutional Political Economy* 13, no. 3 (2002): 284. To reiterate, *catallaxy* requires acceptance

of the naturalistic fallacy: the erroneous view that a desired social state can be comprehensively explained reductively in terms of unobserved natural properties or forces. See Block and Somers, *The Power of Market Fundamentalism*, 38.

62. Friedrich Hayek made the point in 1948, F. Hayek, *The Constitution of Liberty* (Chicago: University of Chicago Press, 2011), 123; Irving Kristol, "Capitalism, Socialism, and Nihilism," *Public Interest*, 31 (1973), 9.

63. Kristol, "Capitalism, Socialism, and Nihilism," 8.

64. Quotations in this paragraph are from ibid., 5, 9.

65. Kristol, "Capitalism, Socialism, and Nihilism," 6. Kristol continues, "I think it is becoming increasingly clear that religion, and a moral philosophy associated with religion, is far more important politically than the philosophy of liberal individualism admits" (12). For critical and sympathetic accounts of Kristol's inability to get beyond the Hayekian paradox, see respectively David Hancock, "Neoconservatism, Bohemia and the Moral Economy of Neoliberalism," *Journal for Cultural Research*, 20 (2016): 101–21; and Michael D. Tanner, *Leviathan on the Right: How Big Government Conservatism Brought Down the Republican Revolution* (Washington, DC: Cato Institute, 2007), 57.

66. On efforts by various New Right figures to overcome the paradox, see Chaves, *Reviewing Political Criticism*. See also, Daniel Bell, *The Coming of Post-industrial Society* (New York: Basic Books, 1973); Bell, *The Cultural Contradictions of Capitalism*, 2nd ed. (London: Heinemann, 1978). The influence on right-leaning politicians of Jude Wanniski's "Taxes and a Two-Santa Theory," *National Observer*, March 6, 1976, is discussed in detail by Christopher G. Faricy, *Welfare for the Wealthy: Parties, Social Spending, and Inequality in the United States* (Cambridge: Cambridge University Press, 2015), 124. The influence of Laffer's "curve" and Gilder's ideas about the role of wealth creation in promoting growth is discussed in John Kenneth Galbraith, *Money: Whence It Came, Where It Went* (Boston: Houghton Mifflin, 1975), 231–32.

67. Hancock, "Neoconservatism," 115–16.

68. Quotations in this paragraph are from ibid., 116; George Gilder, *Wealth and Poverty: A New Edition for the Twenty-First Century* (Washington, DC: Regnery, 2012), 47.

69. Indeed, Gilder would maintain this line of argument even in the wake of the taxpayer funded bank bailouts of 2008, writing in 2012:

> The reason for the huge wealth gap between Larry Page and Suzie Saintly, Donald Trump and Harry Homeless, between Oprah and Obama, or between Eric Schmidt and inventor Dan Bricklin, or the one percent and any number of other worthy men and women is entrepreneurial knowledge and commitment.

See Gilder, *Wealth and Poverty, New Ed.*, xxxv.

70. Phillips-Fein, "Conservatism," 734.

71. Davis, "Late-Imperial America," 30–31.

72. See Kate Bowler, *Blessed: A History of the American Prosperity Gospel* (Oxford: Oxford University Press, 2013); and Colin Campbell, *The Romantic Ethic and the Spirit of Consumerism* (Oxford: Blackwell, 1987).

73. Piven, *Challenging Authority*, 9.

Chapter 4

1. On Bush's environmental promises, see Turner, "The Specter of Environmentalism," 145. Admiration for Bush's compassion and environmentalism is expressed by Jon Meacham, Destiny and Power: The American Odyssey of George Herbert Walker Bush (New York: Random House, 2015), 268, also 471–79. On the Bush administration actually favoring fossil fuel interests over environmentalist demands, and portrayals of such demands as evidence for radical ambitions, see David Schlosberg and Sara Rinfret, "Ecological Modernisation, American Style," Environmental Politics 17, no. 2 (2008): 257. On Clinton's loose ties to the 1960's New Left, see See Jim Mann and Jonathan Peterson, "Anti-War Activists Call Clinton Role Peripheral," Los Angeles Times, October 10, 1992, http://articles.latimes.com/1992-10-10/news/mn-689_1_vietnam-war-protest

2. The academic pejorative left neoliberalism would not gain popular currency until after the global financial crisis of 2007–2008. On Clinton-Gore campaign promises, see Robert Pollin, *Contours of Descent: US Economic Fractures and the Landscape of Global Austerity*, new ed. (London: Verso, 2005), 173–74; also, Gottlieb, *Forcing the Spring*, 3–4; Rosenbaum, *Environmental Politics and Policy*, 13; Norman J. Vig, "Presidential Powers and Environmental Policy," in Vig and Kraft, *Environmental Policy*, 91–92. On radical's hopes, see Phillip F. Cramer, *Deep Environmental Politics: The Role of Radical Environmentalism in Crafting American Environmental Policy* (Westport, CT: Praeger, 1998), 47ff. The influence of the Watergate babies is discussed in Matt Stoller, "How Democrats Killed Their Populist Soul," *Atlantic*, October 24, 2016, http://www.theatlantic.com/politics/archive/2016/10/how-democrats-killed-their-populist-soul/504710/. See also, L. Geismer, *Don't Blame Us: Suburban Liberals and the Transformation of the Democratic Party* (Princeton: Princeton University Press, 2015), 270–73.

3. On shift from government to governance, see James N. Rosenau and Ernst-Otto Czempiel, eds., *Governance without Government: Order and Change in World Politics* (Cambridge: Cambridge University Press, 1992); R. A. W. Rhodes, *Understanding Governance: Policy Networks, Governance, Reflexivity, and Accountability* (Buckingham, UK: Open University Press, 1997); Jon Pierre, ed. *Debating Governance: Authority, Steering, and Democracy* (Oxford: Oxford University Press, 2000); Jon Pierre and B. Guy Peters, *Governance, Politics, and the State*

(London: Macmillan, 2000); and David Held, "At the Global Crossroads: The End of the Washington Consensus and the Rise of Global Social Democracy?," *Globalizations* 2, no. 1 (2005), 95–113.

On the Mandate for Change as blueprint, see Will Marshall and Martin Schram, eds., *Mandate for Change* (Washington, DC: Progressive Policy Institute, 1992); xvi, 198. See also Lily Geismer, *Don't Blame Us: Suburban Liberals and the Transformation of the Democratic Party* (Princeton, NJ: Princeton University Press, 2015), 150, 270ff. On the electoral bridge strategy, see Pollin, *Contours of Descent*, 26 (Clinton quoted at 173).

Arguably, Italian Euro-communists initiated the idea of a postindustrial third way that would allow electoral left parties to incorporate right-wing policies in the late 1960s. For a contemporaneous account, see Norberto Bobbio, *Left and Right: The Significance of a Political Distinction*, trans. Allan Cameron (Cambridge, UK: Polity, 1996). On the third way in the United States, see Stephen Skowronek, *The Politics Presidents Make: Leadership from John Adams to George Bush* (Cambridge, MA: Belknap Press of Harvard University Press, 1997). On the global roots of the New Democrats' third way, see Andrew Scott, *Running on Empty: The "Modernising" of the British and Australian Labour Parties* (Sydney: Pluto, 2000), 4, 118, 163; Thomas Frank, *Listen, Liberal; or, What Ever Happened to the Party of the People?* (New York: Metropolitian Books), 51; and Curtis Atkins, "The Third Way International," *Jacobin*, February 11, 2016.

4. Scerri, *Greening Citizenship*, 101–2.

5. Ibid.; also Atkins, "The Third Way International"; Frank, *Listen Liberal*; Geismer, *Don't Blame Us*.

6. Lily Geismer, "Atari Democrats," *Jacobin*, February 8, 2016, 43.

7. Lester C. Thurow, *The Zero-Sum Society: Distribution and the Possibilities for Economic Change* (New York: Basic Books, 1980); Barry Bluestone and Bennett Harrison, *The Deindustrialization of America: Plant Closings, Community Abandonment, and the Dismantling of Basic Industry* (New York: Basic Books, 1982); and Hazel Henderson, *The Politics of the Solar Age: Alternatives to Economics* (Garden City, NY: Anchor/Doubleday, 1981), all cited in Timothy W. Luke, "Informationalism and Ecology," 59, 64.

8. Passage cited is from "Informationalism and Ecology," 65–66 (italics removed). See also, Piccone, "The Crisis of One-Dimensionality," *Telos*, no. 35 (1978): 43–54; Luke, "Culture and Politics in the Age of Artificial Negativity"; and Luke, *Social Theory and Modernity*.

9. On the administration's liberal stances, see Geismer, "Atari Democrats," 47. On the aspiration to develop an environmental-economic symbiosis, see Martin A. Nie, "'It's the Environment, Stupid!' Clinton and the Environment," *Presidential Studies Quarterly* 27, no. 1 (1997): 39.

10. Vig and Kraft, *Environmental Policy*, 16. A similar point is made in Vig, "Presidential Powers," 92; and Gottlieb, *Forcing the Spring*, 390.

11. For detailed accounts of ecological modernization, see John Barry, "Ecological Modernization," in *Debating the Earth: The Environmental Politics Reader*, ed. John S. Dryzek and David Schlosberg (Oxford: Oxford University Press, 2005), 303–22; Dana R. Fisher and William R. Freudenberg, "Ecological Modernisation and Its Critics: Assessing the Past and Looking towards the Future," *Society and Natural Resources* 14, no. 8 (2001): 701–9; Martin Jänicke, *Ecological Modernization: Innovation and Diffusion of Policy and Technology* (Berlin: Forschungsstelle für Umweltpolitik, Freie Universität Berlin, 2008); Joseph Murphy, "Ecological Modernisation," *Geoforum* 31, no. 1 (2000): 1–8; and Al Gore, *Earth in the Balance: Ecology and the Human Spirit* (New York: Houghton Mifflin, 1992). Gore in particular is enthusiastic about the transformation in human spirit from one adhering to the dualistic view of human's rightful rational mastery to one that holistically conceives of humanity as but one participant in the ecosphere. See Timothy W. Luke, "The Politics of True Convenience or Inconvenient Truth: Struggles over How to Sustain Capitalism, Democracy, and Ecology in the 21st Century," *Environment and Planning A* 40, no. 8 (2008), 1811–15.

12. On the uses of eco-modernization as a political-cultural tool, notably in Germany, the Netherlands, and Japan, see Maarten Hajer, *The Politics of Environmental Discourse: Ecological Modernization and the Policy Process* (Oxford: Clarendon Press, 1995), 31–35. On the role played by eco-modernization nationally and globally, see D. L. Levy and D. Egan, "A Neo-Gramscian Approach to Corporate Political Strategy: Conflict and Accommodation in the Climate Change Negotiations," *Management Studies* 40, no. 4 (2003).

13. On wholesale ecologization, see Albert Weale, *The New Politics of Pollution* (Manchester: Manchester University Press, 1992), 32. On the (re) incorporation of externalities, see Peter Christoff, "Ecological Modernisation, Ecological Modernities," *Environmental Politics* 5, no. 3 (1996): 495; and supermodernity and the prioritization of ecological goals, Christoff, "Out of Chaos, a Shining Star? Toward a Typology of Green States," in *The State and the Global Ecological Crisis*, ed. John Barry and Robyn Eckersley (Cambridge, MA: MIT Press, 2005), 43. On improving citizenly capacities, see Scerri, *Greening Citizenship*, 110.

14. David Schlosberg and John S. Dryzek, "Political Strategies of American Environmentalism: Inclusion and Beyond," *Society and Natural Resources* 15, no. 9 (2002): 795, 800n3.

15. James Meadowcroft, "From Welfare State to Ecostate," in *The State and the Global Ecological Crisis*, ed. John Barry and Robyn Eckersley (Cambridge, MA: MIT Press, 2005), 11.

16. For social-democratic and moral liberal interpretations of this same phenomenon, see Theda Skocpol, *Diminished Democracy: From Membership to Management in American Civic Life* (Norman: University of Oklahoma Press, 2003); and Robert D. Putnam, *Bowling Alone: The Collapse and Revival of American Community* (New York: Simon and Schuster, 2000).

17. Scerri, *Greening Citizenship*, 302–3.

18. Nie, "It's the Environment, Stupid!," 41, 47.

19. For an overview of the literature on the Contract with America that provided the rhetorical spearhead for the Republicans' campaign, see Phillips-Fein, "Conservatism," 726. On congressional demands for a moratorium see, Vig, "Presidential Powers," 92–93. On the failures of the President's Council, see Schlosberg and Rinfret, "Ecological Modernisation, American Style," 258.

20. See, Jay G. Blumler, and Dennis Kavanagh, "The Third Age of Political Communication: Influences and Features," *Political Communication* 16, no. 3 (1999): 209–30; Catherine Needham, "Brand Leaders: Clinton, Blair and the Limitations of the Permanent Campaign," *Political Studies* 53, no. 2 (2005), 343–61. Compare Clinton's 1991 announcement of a New Covenant as the party's central campaign theme with his 1995 State of the Union address, http://www.4president.org/speeches/billclinton1992announcement.htm; https://www.c-span.org/video/?62882-1/president-bill-clintons-1995-state-union-address&start=1374. Perhaps the most well-known resignation was that of Robert Reich as labor secretary. See Reich, *Locked in the Cabinet* (New York: Knopf, 1997).

21. Sheila M. Cavanagh, Robert W. Hahn, and Robert N. Stavins, *National Environmental Policy During the Clinton Years* (Washington, DC: Resources for the Future, 2001), ii.

22. Premilla Nadasen, "How a Democrat Killed Welfare," *Jacobin*, February 9, 2016. See also Premilla Nadasen, Jennifer Mittelstadt, and Marisa Chappell, *Welfare in the United States: A History with Documents, 1935–1996* (New York: Routledge, 2009), ch. 4. While the extent to which Clinton internalized such justifications at the time is an open question, by 2016 he certainly appears to have done so. See Alan Yuhas, "Protesters Put Bill Clinton on the Defensive at Hilary Rally in Philadelphia," *Guardian*, April 7, 2016, https://www.theguardian.com/us-news/2016/apr/07/bill-clinton-protesters-hillary-rally-philadelphia-incarceration. For the composition of state legislatures since 1978, see "Partisan Composition of State Legislatures," *Fact Tank*, Pew Research Center, March 2, 2015, http://www.pewresearch.org/fact-tank/2015/03/02/ahead-of-redistricting-democrats-seek-to-reverse-statehouse-declines/ft_15-03-02_legislatures-1/.

23. Michelle Alexander, "Black Lives Shattered," *The Nation*, February 29, 2016, 13; and *The New Jim Crow: Mass Incarceration in the Age of Colorblindness* (New York: New Press, 2010), 57. See also Naomi Murakawa, *The First Civil Right: How Liberals Built Prison America* (Oxford: Oxford University Press, 2014).

24. James Forman Jr. shows that some of the most ardent supporters of Clinton's tough-on-crime policies were black politicians and community leaders, many veterans of civil rights campaigns. Such policies were not in his view explicitly designed to subordinate African-Americans, but to protect morally

worthy and morally failing African-Americans from the crime and violence plaguing the community. Forman, *Locking up Our Own: Crime and Punishment in Black America* (New York: Farrar, Straus and Giroux, 2017).

25. Nadasen, "How a Democrat Killed Welfare," 61.

26. James Meek, "Robin Hood in a Time of Austerity," *London Review of Books*, February 18, 2016. Recently, it has been reported that in the lead up to the 2016 presidential election, "in a speech at an NAACP convention in Philadelphia in July [2016], Clinton acknowledged that tougher incarceration provisions in the bill were a mistake. 'I signed a bill that made the problem worse. . . . And I want to admit it.'" See Robert Farley, "Bill Clinton and the 1994 Crime Bill," *The Wire* (blog), FactCheck.org, a Project of the Annenburg Public Policy Center, April 12, 2016, http://www.factcheck.org/2016/04/bill-clinton-and-the-1994-crime-bill/.

27. See Chris Rojek, "The Post-Auratic President," *American Behavioral Scientist* 46, no. 4 (2002): 487–500. More recently, see Daryl A. Carter, *Brother Bill: President Clinton and the Politics of Race and Class* (Fayetteville: University of Arkansas Press, 2017).

28. John S. Dryzek, "Political and Ecological Communication." *Environmental Politics* 4, no. 4 (1995): 13–30.

29. Beck, *Risk Society*; *Ecological Politics in an Age of Risk*, trans. Amos Weisz (Cambridge, UK: Polity, 1995); *Ecological Enlightenment: Essays on the Politics of the Risk Society*, trans. Mark A. Ritter (Atlantic Highlands, NJ: Humanities Press, 1995); and *The Reinvention of Politics: Rethinking Modernity in the Global Social Order*, trans. Mark Ritter (Cambridge, UK: Polity, 1997).

30. Christian Hunold and John S. Dryzek, "Green Political Strategy and the State: Combining Political Theory and Comparative History," in *The State and the Global Ecological Crisis*, ed. John Barry and Robyn Eckersley (Cambridge, MA: MIT Press, 2005), 85.

31. Ibid., 88. On radical green-left movements' dependence on state support for eco-modernization, see also Cramer, *Deep Environmental Politics*.

32. See Dryzek, *Deliberative Democracy and Beyond*, 166. See also Dryzek, *Deliberative Global Politics*; "Democratization as Deliberative Capacity Building"; *Foundations and Frontiers of Deliberative Governance*; and *Politics of the Earth*.

33. Brulle, *Agency, Democracy, and Nature*, 280.

34. Brulle, *Democracy and Nature*, 276; Dryzek, *Deliberative Democracy and Beyond*, 111; Dryzek, *Deliberative Global Politics*, 140; and Dryzek, *Politics of the Earth*, 233–40. 35. Schlosberg and Dryzek, "Political Strategies," 795; see also, Dryzek et al., *Green States*, 133.

36. See George W. Bush, "State of the Union Address," White House, January 31, 2006: https://georgewbush-whitehouse.archives.gov/stateoftheunion/2006/

37. Vig and Kraft, *Environmental Policy*, 17–19; a similar point is made in Vig, "Presidential Powers," 94–98; and Rosenbaum, *Environmental Politics and Policy*, 13–14.

38. Oddly, the post-environmentalists promoted a kind of individualistic green utilitarianism over any kind of collective green political agenda on the basis of this argument, eventually calling for a new Apollo project that would treat environmental problems in the same way as had the cold war nation treated the race to the moon. See Michael Shellenberger and Ted Nordhaus, *The Death of Environmentalism: Global Warming Politics in a Post-environmental World* (Washington, DC: Environmental Grantmakers Association / The Breakthrough Institute, 2004).

39. Zygmunt Bauman, *The Individualized Society* (Cambridge, UK: Polity, 2001); and Richard Sennett, *The Culture of the New Capitalism* (London: Yale University Press, 2005).

40. Scerri, *Greening Citizenship*, 108.

41. Bob Jessop refers to "electoral neocommunitarianism": "Liberalism, Neoliberalism, and Urban Governance: A State-Theoretical Perspective," *Antipode* 34, no. 3 (2002): 463. See also Neil Brenner, *New State Spaces: Urban Governance and the Rescaling of Statehood* (Oxford: Oxford University Press, 2004); Neil Brenner and Nik Theodore, eds., *Spaces of Neoliberalism: Urban Restructuring in North America and Western Europe* (London: Wiley-Blackwell, 2003); and William Davies, "The Emerging Neocommunitarianism," *Political Quarterly* 83, no. 4 (2012): 767–76.

42. Luke, "The System of Sustainable Degradation," *Capitalism, Nature, Socialism* 17, no. 1 (2006): 99–112.

43. See Colin Crouch, *Post-democracy* (Cambridge, UK: Polity, 2004); Crouch, *The Strange Non-death of Neoliberalism* (Cambridge, UK: Polity, 2011), 114ff.; Wolfgang Streeck, *Buying Time: The Delayed Crisis of Democratic Capitalism*, trans. Patrick Camiller (London: Verso, 2014); Streek's disparagement of the "marijuana left" can be found in an interview at https://www.theguardian.com/books/2016/dec/09/wolfgang-streeck-the-german-economist-calling-time-on-capitalism; Stephen Gill, "New Constitutionalism, Democratisation and Global Political Economy," *Pacifica Review* 10, no. 1 (1998): 23–38; Gill, "Market Civilization, New Constitutionalism and World Order," in *New Constitutionalism and World Order*, ed. Stephen Gill and A. Claire Cutler (Cambridge: Cambridge University Press, 2015), 29–44; and, see also Philip Mirowski, *Never Let a Serious Crisis Go to Waste: How Neoliberalism Survived the Financial Meltdown* (London: Verso, 2013).

44. Chantal Mouffe, *The Return of the Political* (London: Verso, 1993); *The Democratic Paradox* (London: Verso, 2000); *On the Political* (London: Routledge, 2005); Wendy Brown, *Edgework: Critical Essays on Knowledge and Politics* (Princeton, NJ: Princeton University Press, 2005); *Undoing the Demos: Neoliberalism's Stealth Revolution* (New York: Zone Books, 2015); Nancy Fraser, "Legitimation Crisis?" 158–59.

45. See Blühdorn, in relation to the Western European experience, *Post-Ecologist Politics*; "Self-Experience in the Theme Park of Radical Action?

Social Movements and Political Articulation in the Late-Modern Condition," *European Journal of Social Theory* 9, no. 1 (2006): 29–30; and "The Governance of Unsustainability: Ecology and Democracy after the Post-democratic Turn," *Environmental Politics* 22, no. 1 (2013): 18–21.

46. See MacLean, *Democracy In Chains*; S. M. Amadae, *Rationalizing Capitalist Democracy*; and *Prisoners of Reason: Game Theory and Neoliberal Political Economy* (New York: Cambridge University Press, 2016).

46. The Pew Research Center conducts multiple, often longitudinal, surveys. Of particular interest is that addressing "Political Polarization, 1994–2017," http://www.people-press.org/interactives/political-polarization-1994-2017/.

47. Possibly first described by Jean-Francois Lyotard, *The Postmodern Condition: A Report on Knowledge*, trans. Geoff Bennington and Brian Massumi (Minneapolis: University of Minnesota Press, 1993), 1. More recently, see David Runciman, *The Confidence Trap: A History of Democracy in Crisis from World War I to the Present* (Princeton, NJ: Princeton University Press, 2013), 331–36.

Chapter 5

1. The economic concept of market externality was first used in political science in the early 1970s. See Thomas C. Schelling, "Hockey Helmets, Concealed Weapons, and Daylight Saving: A Study of Binary Choices with Externalities," *Journal of Conflict Resolution* 17, no. 3 (1973): 381–428. Indeed, by the time of the protests against the World Economic Forum in Seattle in 1999, the business lobby was portraying the new New Left as both a genuine threat to American free-enterprise and as an irrelevance to the Washington consensus as the only show in town. See Noah Smith, "The Dark Side of Globalization: Why Seattle's 1999 Protesters Were Right," *Atlantic* January 6, 2014, http://www.theatlantic.com/business/archive/2014/01/the-dark-side-of-globalization-why-seattles-1999-protesters-were-right/282831/; and Anonymous, "The New Trade War: Opponents of Globalisation Wreaked Havoc at a Big Trade Summit in Seattle This Week," *Economist*, December 2, 1999. And James K. Rowe, "Corporate Social Responsibility as Business Strategy," in *Globalization, Governmentality and Global Politics: Regulation for the Rest of Us?* ed. Ronnie D. Lipschutz and James K. Rowe (London: Routledge, 2005), 141.

2. For criticism of the radical externalization thesis as an example of neoliberal ideology in action, see Ronen Shamir, "The Age of Responsibilization: On Market-Embedded Morality," *Economy and Society* 37, no. 1 (2008): 3–4; see also, Crouch, *The Strange Non-death*, 129, 138.

3. Beck, *Risk Society*; Beck, *Ecological Politics*; and Brian Tokar, *Toward Climate Justice: Perspectives on the Climate Crisis and Social Change*, rev. ed. (Porsgrunn, Norway: New Compass, 2014), 74. See also Naomi Klein, *No Logo*

(New York: Picador, 1999); and Joel Bakan, *The Corporation: The Pathological Pursuit of Profit and Power* (Toronto: Viking Canada, 2004).

4. Examples include Philip Selznick, *The Communitarian Persuasion* (Baltimore: Johns Hopkins University Press, 2002), 101; Simon Zadek, *The Civil Corporation: The New Economy of Corporate Citizenship* (London: Earthscan, 2001); Stuart L. Hart, "Beyond Greening: Strategies for a Sustainable World," *Harvard Business Review* 75, no. 1 (1997): 66–76; Christine Parker, *The Open Corporation: Effective Self-Regulation and Democracy* (Cambridge: Cambridge University Press, 2002); David Vogel, *The Market for Virtue: The Potential and Limits of Corporate Social Responsibility* (Washington, DC: Brookings Institution Press, 2005); Fritjof Capra, *The Tao of Physics: An Exploration of the Parallels between Modern Physics and Eastern Mysticism* (Berkeley, CA: Shambhala, 1975); Fritjof Capra and Pier Luigi Luisi, *The Systems View of Life: A Unifying Vision* (Cambridge: Cambridge University Press, 2014); Marc Benioff and Karen Southwick, *Compassionate Capitalism: How Corporations Can Make Doing Good an Integral Part of Doing Well* (Franklin Lakes, NJ: Career Press, 2003); Paul Hawken, Amory Lovins, and L. Hunter Lovins, *Natural Capitalism: Creating the Next Industrial Revolution* (Boston: Back Bay Books, 2000); and Brian Nattrass and Mary Altomare, *The Natural Step for Business: Wealth, Ecology, and the Evolutionary Corporation* (Gabriola Island, BC: New Society, 1999).

5. The original argument for corporate social responsibility is probably found in Howard R. Bowen, *The Social Responsibilities of the Businessman* (New York: Harper and Row, 1953). Friedman asserted this claim in the early 1970s. See Milton Friedman, "The Social Responsibility of Business Is to Increase Profits," *The New York Times Magazine*, Sep. 13 1970. The countercultural roots of the movement are allued to in Vogel, *The Market for Virtue*; Fritjof Capra, *The Turning Point: Science, Society, and the Rising Culture* (New York: Bantam, 1983); Paul Hawken, *Next Economy* (New York: Ballantine, 1983). On the unhelpfulness of the "enemy perception" advocated by radicals, see Peter Knorringa and A. H. J. Helmsing, "Beyond an Enemy Perception: Unpacking and Engaging the Private Sector," *Development and Change* 39, no. 6 (2008): 1053–62. For a critique of the corporate responsibility movement's political blind spots or omissions, see, James K. Rowe, "Corporate Social Responsibility as Business Strategy," *Globalization, Governmentality and Global Politics: Regulation for the Rest of Us?* eds. Ronnie Lipschut and James K. Rowe (London: Routledge, 2005), 130–70.

6. At the Rio de Janeiro Earth Summit in 2012, the United Nations General Assembly published its report and webpage, "The Future We Want." The declaration makes explicit the call for the "private sector to engage in responsible business practices" and for support for the "integration of the three dimensions of sustainable development in a holistic and cross-sectoral manner at all levels." See, United Nations General Assembly, *The Future We Want*, New York: United Nations General Assembly, 2012, available at https://sustainable

development.un.org/futurewewant.html. Indeed, the ideas and practices associated with corporate social and environmental responsibility rapidly emerged as the defining ideas and practices within the corporate sector. As such, a selection of illustrative empirical material suffices to make the point that the movement does indeed represent a historical-epochal shift. Each of the four major global corporate social responsibility umbrella groups—AccountAbility International, Global Reporting Initiative, International Standards Organization, and the United Nations Global Compact—refer directly to the failures of dualistic regulatory institutions and their inability to foster responsible, ethical, innovative, and creative solutions and do so in holistic terms. Major multinational corporations working across a range of industries also commit themselves to holistic ecology based on claims that the old ways of doing things are outmoded, inefficient, and unsustainable. A selection of "responsible" firms that make holistic ecological claims includes conglomerates such as Unilever, United Parcel Service, and Coca-Cola; information technology firms such as Advanced Micro Devices, Microsoft, and Dell; and pharmaceutical corporations Merck, Novo Nordisk, and Sanofi. Internationally, the United Nations Environment Program–sponsored Economics of Ecosystems and Biodiversity is explicitly committed to promoting holistic ecology and market-centered supply-side policies. More interestingly, the World Bank supports the Initiative for Sustainable Forest Landscapes and Center for International Forestry Research, which seeks to reorient existing Reducing Emissions from Deforestation and forest Degradation (REDD) schemes around a holistic "landscapes approach." See, AccountAbility, *Accountability—Trends in Corporate Responsibility Reporting*, Institute of Social and Ethical Accountability, http://www.accountability.org/advisory-services/strategy-and-governance/; Global Reroprting Iinitiative, "Dutch Decade 2: In Conversation with Jan-Peter Balkenende on National Sustainability Day," News and Press Center, GRI website, October 10, 2012, https://www.globalreporting.org/information/news-and-press-center/Pages/Dutch-decade-2-in-conversation-with-Jan-Peter-Balkenende-on-national-sustainability-day.aspx; GRI, UNEP, KPMG, UCGA, "Carrots and Sticks—Promoting Transparency and Sustainability: An Update on Trends in Voluntary and Mandatory Approaches to Sustainability Reporting," (Nairobi: Global Reporting Initiative/United Nations Environment Programme/KPMG/Unit for Corporate Governance in Africa, 2009); United Nations Global Compact, "Address Environmental Risks and Leverage Opportunities," https://www.unglobalcompact.org/what-is-gc/our-work/environment; TEEB, *The Economics of Ecosystems and Biodiversity Ecological and Economic Foundations*, ed. Pushpam Kumar (London and Washington: Earthscan, 2010); Steve Zwick, "Unpacking Warsaw, Part Two: Recognizing the Landscape Reality," news articles, Ecosystem Marketplace, December 5, 2013, http://www.ecosystemmarketplace.com/articles/em-unpacking-warsaw-part-two-em-recognizing-the-landscape-reality/; Coca Cola, *2012/2013 GRI Report* (Atlanta: Coca Cola Company, 2012); Advanced Micro

Devices, *2012/2013 Corporate Responsibility Report* (Sunnyvale, CA: AMD, 2013); Dell Corporation, *FY13 Corporate Responsibility Summary Report* (Plano, TX: Dell Corporation, 2013); Merck, "Corporate Responsilbility Highlights 2012," (Whitehouse Station, NJ: Office of Corporate Responsibility, Merck, 2013); Microsoft Inc., *Citizenship Report 2013* (Seattle: Microsoft, 2013); Novo Nordisk, *2013 Annual Report* (Bagsvaerd, Denmark: Novo Nordisk, 2013); Sanofi, *Corporate Social Responsibility Report 2012* (Paris: Sanofi, 2012); and United Parcel Service, "Global Volunteer Week," http://www.community.ups.com/volunteerism/gvw.html.

7. Brown, *Undoing the Demos*, 25; and Rosenbaum, *Environmental Politics and Policy*, 85.

8. Luke, "System of Sustainable Degradation," 103ff. For similar criticisms, see Christopher Wright and Daniel Nyberg, *Climate Change, Capitalism, and Corporations: Processes of Creative Self-Destruction* (Cambridge: Cambridge University Press, 2015); Subhabrata Bobby Banerjee, *Corporate Social Responsibility: The Good, the Bad and the Ugly* (Northampton, MA: Edward Elgar, 2007).

9. Ronen Shamir, "Capitalism, Governance, and Authority: The Case of Corporate Social Responsibility," *Annual Review of Law and Social Science* 6 (2010): 540, citing Doreen McBarnet, Aurora Voiculescu, and Tom Campbell, eds., *The New Corporate Accountability: Corporate Social Responsibility and the Law* (Cambridge: Cambridge University Press, 2007).

10. Ronen Shamir, "The De-radicalization of Corporate Social Responsibility," *Critical Sociology* 30, no. 3 (2004): 675–76 and "Capitalism, Governance, and Authority: The Case of Corporate Social Responsibility," 540. Eve Chiapello makes a similar argument. See Eve Chiapello, "Capitalism and Its Criticisms," in *New Spirits of Capitalism? Crises, Justifications, and Dynamics*, ed. Paul du Gay and Glenn Morgan (Oxford: Oxford University Press, 2013), 77–78.

11. Shamir, "Capitalism, Governance, and Authority," 534. See also Chiapello, "Capitalism and Its Criticisms," 61.

12. Blühdorn, "The Governance of Unsustainability," 32n5.

13. Dryzek's efforts to rescue communicative rationality from Habermas outlined in John S. Dryzek, "Political and ecological communication," *Environmental Politics* 4, no. 4 (1995): 20. The potential coerciveness of real-world instantiations of Habermas' discourse ethics were possibly first noted by feminist readers of Habermas's critical theory, notably Nancy Fraser, *Unruly Practices*; and "Rethinking the Public Sphere: A Contribution to the Critique of Actually Existing Democracy," in *Habermas and the Public Sphere*, ed. Craig Calhoun (Cambridge, MA: MIT Press, 1992), 109–42.

14. Scerri, "Deep Ecology, the Holistic Critique of Enlightenment Dualism, and the Irony of History," *Environmental Values* 25, no. 5 (2016): 543.

15. Timothy W. Luke, *Ecocritique: Contesting the Politics of Nature* (Minneapolis: University of Minnesota Press, 1997); "Environmentality as Green Governmentality"; *Capitalism, Democracy, Ecology*; and "Corporate Social

Responsibility: An Uneasy Merger of Sustainability and Development," *Sustainable Development* 21, no. 2 (2013): 83–84.

16. Interestingly, Luke does leave room for "developing and nurturing an organic negativity." Such organic negativity would offer genuine critique of the green governmentality regime not only from without the system but also from within the margins of academia, the info-tech professions, or cultural industries, for example. See, *Social Theory and Modernity*, 12, 159–61.

17. Luke, "System of Sustainable Degradation," 111–12.

18. For Luke, like the gestures of those administering the governance turn from Washington, those of business advocates of corporate, social, and environmental responsibility and green-growth capitalism are "neither cynical nor symbolic." See "Corporate Social Responsibility," 88. See also "Developing Planetarian Accountancy: Fabricating Nature as Stock, Service, and System for Green Governmentality," in *Nature, Knowledge and Negation*, ed. Harry F. Dahms (Bingley, UK: Emerald, 2009): 129–59. In addition, see similar arguments by Eva Lövbrand, Johannes Stripple, and Bo Wiman, "Earth System Governmentality: Reflections on Science in the Anthropocene," *Global Environmental Change* 19, no. 1 (2009): 7–13; Stephanie Rutherford, "Green Governmentality: Insights and Opportunities in the Study of Nature's Rule," *Progress in Human Geography* 31, no. 3 (2007): 291–307; Éric Darier, ed., *Discourses of the Environment* (Malden, MA: Blackwell, 1999); and Arun Agrawal, *Environmentality: Technologies of Government and the Making of Subjects* (Durham, NC: Duke University Press, 2005).

19. John M. Meyer defines "dualist" and "derivative" as the two competing conceptions of nature in Western political culture: see *Political Nature*. See note 11, chapter 1; Luke, "Corporate Social Responsibility," 89.

20. Horkheimer and Adorno, *Dialectic of Enlightenment*, 148ff.; Held, *Introduction to Critical Theory*, 155. 24.

21. Orly Lobel cites seminal legal pluralist arguments by Harold Laski on the false enlightenment assumption, and paraphrases the words of Mark Dewolfe Howe on the rights of private groups. See Orly Lobel, "The Paradox of Extralegal Activism: Critical Legal Consciousness and Transformative Politics," *Harvard Law Review* 120, no. 4 (2007), 967; Harold Laski, *Authority in the Modern State* (New Haven, CT: Yale University Press, 1919), 26–27; Mark Dewolfe Howe, *The Supreme Court, 1952 Term—Foreword: Political Theory and the Nature of Liberty, Harvard Law Review* 67, no. 91 (1953).

22. For a detailed account of the relationship between advocates of public choice theory, law and economics, and the New Right response to the "crisis of democracy," see MacLean, *Democracy in Chains*, 115. Maclean often relies on the legal history of Steven M. Teles, *The Rise of the Conservative Legal Movement: The Battle for Control of the Law* (Princeton, NJ: Princeton University Press, 2008).

23. Alain Supiot, *Homo Juridicus: On the Anthropological Function of the Law*, trans. Saskia Brown (London: Verso, 2007): xxi. Wendy Brown in particular discusses "neoliberal jurisprudence," *Undoing the Demos*, 151ff. See also David Singh Grewal and Jedediah Purdy, "Introduction: Law and Neoliberalism," *Law and Contemporary Problems* 77, no. 4 (2014): 1–24.

24. Quotation on defining justice, from Amadae, *Prisoners of Reason*, 205ff.; and on the idea that everything works best on market logic, from Jedediah Purdy, "The Roberts Court v. America," *Democracy*, no. 23 (2012): 2.

25. Timothy K. Kuhner, "*Citizens United* as Neoliberal Jurisprudence: The Resurgence of Economic Theory," *Virginia Journal of Social Policy and the Law* 18, no. 3 (2011): 395, 397. Oddly, Kuhner regards federal government legislative intervention as holistic and so aimed at establishing substantive "equality in matters of race, religion, gender, and national origin occupied the collective consciousness and legal landscape." See *Capitalism v. Democracy: Money in Politics and the Free Market Constitution* (Stanford, CA: Stanford Law Books, an imprint of Stanford University Press, 2014), 43. In contrast, as should be clear, I regard holism as in principle opposed to equality: insofar as the logic of the former depends on the principle of hierarchy, the logic of the latter depends on the artificial principle of a social contract to establish formal equality.

26. Amadae, *Prisoners of Reason*, 211.

27. Interestingly, Luke offers a concept of eco-jurisprudence that highlights the tendency of market interests to exploit legal means in order to facilitate favorable decisions. Understood as part of the turn to governance, this exploitation of legal means serves to distance businesses from oversight by legislative authority or to stick a spoke in the wheels of green (and labor) activists also hoping to obtain legislation or to similarly juridicalize environmental (and labor) issues. Luke, "System of Sustainable Degradation," 109.

28. See *Burwell v. Hobby Lobby Stores*, 134 S. Ct. 2751 (2014).

29. Quotations in this paragraph are from Supreme Court of the United States, *Citizens United v. Federal Electoral Commission*, in *558 U.S. ____(2010)*, ed. by Tenth Circuit US Supreme Court (Washington DC: Supreme Court of the United States, 2010), 48–49; *Burwell*; and Brown, *Undoing the Demos*, 155.

30. Powell, "Confidential Memorandum: Attack on American Free Enterprise System," 10.

31. See, Laura Pulido, Ellen Kohl, and Nicole-Marie Cotton, "State Regulation and Environmental Justice: The Need for Strategy Reassessment," *Capitalism, Nature, Socialism* 27, no. 2 (2016): 12–31; Raoul S. Liévanos, "Certainty, Fairness, and Balance: State Resonance and Environmental Justice Policy Implementation," *Sociological Forum* 27, no. 2 (2012): 481–503; Peter Dauvergne and Genevieve LeBaron, *Protest Inc.: The Corporatization of Activism* (Cambridge, UK: Polity, 2014); Peter Dauvergne and Jane Lister, *Eco-business: A Big Brand*

Takeover of Sustainability (Cambridge, MA: MIT Press, 2013); and Ryan Holifield, "Neoliberalism and Environmental Justice Policy," in *Neoliberal Environments: False Promises and Unnatural Consequences*, ed. Nik Heynen, James McCarthy, Scott Prudham, and Paul Robbins (New York: Routledge, 2007), 202–16.

32. Shamir, "Age of Responsibilization," 6.

33. Shamir, "Capitalism, Governance, and Authority," 536.

34. Ibid., 545; and "Age of Responsibilization," 14. Shamir argues that such an ethics is consequentialist. However, this seems to imply that the new jurisprudence is ethically concerned with the consequences of decisions beyond the minimalist purview of the law. Rather, it seems that such jurisprudence is utilitarian insofar as the remit of the law is restricted to ensuring legal persons' primitive freedom to maximize happiness as measured in units of economic utility, as the *Hobby Lobby* decision illustrates. See my discussion of eco-authoritarianism in chapter 2.

35. Andy Scerri and Nader Sobhani, "Even Natural Disasters Are Unlikely to Slow Us Down: Corporate Social & Environmental Responsibility as Well-Crafted Political Judgment," in *Biopolitical Disaster*, eds. Jennifer Lawrence and Sara-Marie Wiebe (London: Routledge, 2018), 62–77.

36. Frank, *Listen, Liberal*; Fraser, *The Limousine Liberal*.

37. Radkau, *Age of Ecology*, 9–10.

38. Naomi Klein, "It's Hard Work Trying to Escape from the Gooey Corporate Hug," *The Guardian Weekly* 14–20 (international edition), June 2001, https://www.theguardian.com/theguardian/2001/jun/14/guardianweekly.guardianweekly11.

39. Radkau, *Age of Ecology*, 8.

Chapter 6

1. Perhaps the most well-known iteration of such triumphalism being that supplied by Francis Fukuyama in *The End of History and the Last Man* (New York: Free Press, 1992). On the emergence of the global justice movement, see Manfred B. Steger, *The Rise of the Global Imaginary: Political Ideologies from the French Revolution to the Global War on Terror* (Oxford: Oxford University Press, 2008); and Manfred B. Steger, James Goodman, and Erin K. Wilson, *Justice Globalism: Ideology, Crises, Policy* (Thousand Oaks, CA: Sage, 2012).

2. See Malcolm Waters, *Globalization* (London: Routledge, 1995); Richard Falk, *Predatory Globalization: A Critique* (Cambridge, UK: Polity, 1999); Amory Starr, *Naming the Enemy: Anti-corporate Movements Confront Globalization* (Sydney: Pluto, 2000); Marianne Maeckelbergh, *The Will of the Many: How the Alterglobalisation Movement Is Changing the Face of Democracy* (London: Pluto, 2011); Geoffrey Pleyers, *Alter-globalization: Becoming Actors in a Global Age* (Cambridge, UK: Polity, 2010); Gilles Deleuze, *Difference and Repetition*, trans. Paul Patton (New York.: Columbia University Press, 1994); Gilles Deleuze and Félix Guattari, *Anti-Oedipus:*

Capitalism and Schizophrenia, trans. Robert Hurley, Mark Seem, and Helen R. Lane (London: Athlone, 1977); Deleuze and Guattari, *A Thousand Plateaus: Capitalism and Schizophrenia*, trans. Brian Massumi (Minneapolis: University of Minnesota Press, 1987); Michael Hardt and Antonio Negri, *Empire* (Cambridge, MA: Harvard University Press, 2000); Hardt and Negri, *Multitude: War and Democracy in the Age of Empire* (London: Penguin, 2005); and Hardt and Negri, *Commonwealth* (Camrbidge, MA: Belknap Press of Harvard University Press, 2009).

For a contemporaneous account of the link between French 1960s Heideggerian Spinozism, notably work by Deleuze and Guattari, and work by Hardt and Negri and the neo-anarchist global justice movements, see Graeme Chesters and Ian Welsh, *Complexity and Social Movements: Multitudes at the Edge of Chaos* (London: Routledge, 2006). And Andrew Cornell explicitly recognizes Hardt and Negri's influence on neo-anarchist prefigurative movements in the United States: see *Oppose and Propose! Lessons from Movement for a New Society* (Oakland, CA: AK Press, 2011), 167–68. For a dispassionate account of Heidegger's Spinozism, see James Luchte, "Of Freedom: Heidegger on Spinoza," *Epoché* 20, no. 1 (2015): 131–47.

Quotations in this paragraph are from Tom Nairn, "Make for the Boondocks," *London Review of Books*, May 5, 2005, 12; and Epstein, *Political Protest and Cultural Revolution*, 243. See also, Richard Rorty, *Achieving our Country: Leftist Thought in Twentieth Century America* (Cambridge MA, Harvard University Press, 1998), pp. 83–87.

3. See David Graeber, *Democracy Project: A History, a Crisis, a Movement* (New York: Spiegel and Grau, 2013), 120–21; and Zerzan, *Emptiness*, 165ff.

4. Alongside Thoreau and Bookchin, influences include William Golding, Peter Kropotkin, Mikhail Bakunin, Pierre-Joseph Proudhon, and Josiah Warren. See David Graeber, *Fragments of an Anarchist Anthropology* (Chicago: Prickly Paradigm Press, 2004); Graeber, *The Democracy Project: A History, a Crisis, a Movement* (New York: Spiegel and Grau, 2013); Simon Critchley, *Infinetely Demanding: Ethics of Commitment, Politics of Resistance* (London: Verso, 2007); Paul Kingsnorth, *One No, Many Yeses: A Journey to the Heart of the Global Resistance Movement* (London: Free Press, 2003); Hakim Bey, *T.A.Z.: The Temporary Autonomous Zone, Ontological Anarchy, Poetic Terrorism* (New York: Autonomedia, 1991); Richard J. F. Day, *Gramsci Is Dead: Anarchist Currents in the Newest Social Movements* (London: Pluto, 2005); John Holloway, *Change the World without Taking Power* (London: Pluto, 2002); The Invisible Committee, *The Coming Insurrection* (Los Angeles: Semiotext(e), 2009); and The Invisible Committee, *To Our Friends*, trans. Robert Hurley (Los Angeles: Semiotext(e), 2014).

5. Graeber, *Fragments of an Anarchist Anthropology*, 3. See also Bey, *T.A.Z*; and John Zerzan, *Running on Emptiness: The Pathology of Civilization* (Port Townsend, WA: Feral House, 2002).

6. For J. K. Gibson-Graham, the revivification of green-left activism in the 1990s in fact brought "new life" to a synthetic "anarchic situationism,"

where the latter is understood to give expression to the critical theory of "social theorists and philosophers like Deleuze and Guattari." See J. K. Gibson-Graham, *A Postcapitalist Politics* (Minneapolis: University of Minnesota Press, 2006), xx, xxxi; Gibson-Graham, *The End of Capitalism (as We Knew It): A Feminist Critique of Political Economy* (Oxford: Blackwell, 2006), esp. 251ff.; and Gerda Roelvink, Kevin St. Martin, and J. K. Gibson-Graham, eds., *Making Other Worlds Possible: Performing Diverse Economies* (Minneapolis: University of Minnesota Press, 2015). Emphases added to highlight the conflation of "personal and social" with "class."

7. Roelvink, St. Martin, and Gibson-Graham, *Making Other Worlds Possible*, 8.

8. David Graeber, "The New Anarchists," *New Left Review*, no. 13 (2002): 68; Graeber, *Fragments of an Anarchist Anthropology*, 3, 65ff.; and Gibson-Graham, *A Postcapitalist Politics*, xxvi.

9. Graeber, *Fragments of an Anarchist Anthropology*, 1, where Kropotkin is cited.

10. Ibid., 73. Compare with Hardt and Negri's musings on the "being in becoming" of the global "multitude."

11. Gibson-Graham, *A Postcapitalist Politics*, 127.

12. Blair Taylor paraphrases Graeber, Simon Critchley, and Richard J. F. Day: see "From Alterglobalization to Occupy Wall Street: Neoanarchism and the New Spirit of the Left," *City* 17, no. 6 (2013): 735.

13. Gibson-Graham, *A Postcapitalist Politics*, xxxiv; and *The End of Capitalism*, ix, 257.

14. Roelvink, St. Martin, and Gibson-Graham, *Making Other Worlds Possible*, 8, 10; and Gibson-Graham, *A Postcapitalist Politics*, xxvi.

15. Luke Yates, "Rethinking Prefiguration: Alternatives, Micropolitics and Goals in Social Movements," *Social Movement Studies* 14, no. 1 (2015): 3.

16. On prefiguration as fulfilling certain conditions, Yates, "Rethinking Prefiguration," 3; examples from Maeckelbergh, *The Will of the Many*, 7; and Sean Parson, "Parks, Permits, and Riot Police: San Francisco Food Not Bombs and Autonomous Occupations of Space," *New Political Science* 37, no. 3 (2015): 346–62.

17. Graeber, "The New Anarchists," 70.

18. Roelvink, St. Martin, and Gibson-Graham, *Making Other Worlds Possible*, 4.

19. Ibid., 7.

20. Ibid., 8. Examples are taken from case studies presented in the text.

21. Gibson-Graham, *A Postcapitalist Politics*, xxx.

22. Taylor, "From Alterglobalization," 733.

23. Ibid., 735.

24. See, in addition to numerous media articles, Simon Critchley, *Infinitely Demanding*; Critchley, *The Faith of the Faithless: Experiments in Political Theology* (London: Verso, 2012); Slavoj Žižek, "Resistance Is Surrender," *London Review*

of *Books*, November 15, 2007; Žižek, *In Defense of Lost Causes* (London: Verso, 2009).

25. Taylor, "From Alterglobalization," 735, 745.

26. Taylor, "From Alterglobalization," 745.

27. Brunkhorst, *Adorno and Critical Theory*, 138, 144; Offe, *Reflections on America*, 91–92; and Mariotti, *Adorno and Democracy*, 5, 28.

28. Nairn, "Make for the Boondocks," 13.

29. Roelvink, St. Martin, and Gibson-Graham, *Making Other Worlds Possible*, 6.

30. Gibson-Graham, *A Postcapitalist Politics*, xxxvii (emphases in original).

31. Graeber, "The New Anarchists," 71–72.

32. Graeber addresses these criticisms somewhat ineffectively in "The New Anarchists" and *Fragments of an Anarchist Anthropology*.

33. See, Jane Bennett, *Vibrant Matter: A Political Ecology of Things* (Durham, NC: Duke University Press, 2010); and "From Nature to Matter," in *Second Nature: Rethinking the Natural through Politics*, ed. Crina Archer, Laura Ephraim, and Lida Maxwell (New York: Fordham, 2013), 9, 154.

34. David Schlosberg and Romand Coles, "The New Environmentalism of Everyday Life: Sustainability, Material Flows and Movements," *Contemporary Political Theory* 15, no. 2 (2016): 168, 173.

35. John M. Meyer and Jens M. Kersten, eds., *The Greening of Everyday Life: Challenging Practices, Imagining Possibilities* (Oxford: Oxford University Press, 2016), 4; and John M. Meyer, *Engaging the Everyday: Environmental Social Criticism and the Resonance Dilemma* (Cambridge, MA: MIT Press, 2015), 64ff., 172–73.

36. Micah White, *The End of Protest: A New Playbook for Revolution* (Toronto: Knopf Canada, 2016), 74–75. Direct quote is from an interview with White in Jaime Lubin, "The Spirit of Activism: From Occupy Wall Street to a New Playbook for Revolution," *The Blog, Huffington Post*, April 27, 2016, http://www.huffingtonpost.com/jaime-lubin/the-spirit-of-activism-fr_b_9780946.html.

37. For a political theoretical delimitation of face-to-face communities as opposed to "agency-extended" (i.e., institution-based) social forms, see James, *Nation Formation*. Often referred to is work by Frans de Waal. See *The Age of Empathy: Nature's Lessons for a Kinder Society* (New York: Harmony Books, 2009).

38. Adorno, *Minima Moralia*, 26; Nathan Kowalsky, "Whatever Happened to Deep Ecology?" *Trumpeter* 30, no. 2 (2014): 95. See Bookchin, "The Communalist Project," *Harbinger* 3, no. 1 (2002): 1–13. See also Tokar, "On Bookchin's Social Ecology," 60–61; Blair Taylor, "Social Ecology in a Neoliberal Age" (unpublished manuscript, 2016); and Murray Bookchin, *The Next Revolution: Popular Assemblies and the Promise of Direct Democracy*, ed. Debbie Bookchin and Blair Taylor (London: Verso, 2015), xxi.

39. As I have argued elsewhere in relation to similar tendencies within the global World Social Forum, "if possibilities for experiencing [freedom] are to be

universalized, the immediate 'eventful' or 'performative' experience of it would need to be subsumed to the . . . tasks of formalizing and proceduralizing, [to] establishing a process" for both engaging and exercising political power. Andy Scerri, "The World Social Forum: Another World Might Be Possible," *Social Movement Studies* 12, no. 1 (2013): 119.

40. Williams, *Beginning*, 82, 92.

41. Epstein, *Political Protest and Cultural Revolution*, 276.

42. Ibid., 275, 276. At around the same time, Gramscian critic of green-left modernist prefigurative movements Carl Boggs argued that

> the "left" presence within some [new social] movements (above all, the ecologists) is commonly linked to a neoanarchism with its utopian vision of the future, its unrelenting critique of bourgeois society, and its holistic commitment to social change. But its strong appeals are undermined by a shallow conception of history—by a romantic attachment to a mythic (often preindustrial) past that coexists with an urgent deisre for immediate and total overthrow of the power structure.

C. Boggs, *Social Movements and Political Power: Emerging Forms of Radicalism in the West* (Philadelphia: Temple, 1986), 14.

43. Epstein, *Political Protest and Cultural Revolution*, 269.

44. See Bookchin, "The Communalist Project," 5.

45. Benjamin J. Pauli, "The New Anarchism in Britain and the US: Towards a Richer Understanding of Post-war Anarchist Thought," *Journal of Political Ideologies* 20, no. 2 (2015): 147.

46. For an indication of Bookchin's growing alienation from "mainstream" anarchist and green-left modernist circles, see Tokar, "On Bookchin's Social Ecology."

47. Andrew Cornell cites calls to anarchists by International Workers of the World organizer Sam Dolgoff in 1964 to acknowledge the failures of the past. See Cornell, *Oppose and Propose!* 173. For a comprehensive history of calls to reconsider the presumptions of prefigurative politics and ahistoricism since the late 1990s, see Chris Dixon, *Another Politics: Talking across Today's Transformative Movements* (Berkeley, CA: University of California Press, 2014), 295n1. Graeber's 2013 newspaper article is cited in Dixon, *Another Politics*, 229. Indeed, Dixon dedicates an entire chapter to "fighting against amnesia."

48. Astra Taylor, "Against Activism," *Baffler*, March 2016.

49. L. A. Kauffman, "The Theology of Consensus," *Berkeley Journal of Sociology*, May 26, 2015, http://berkeleyjournal.org/2015/05/the-theology-of-consensus/.

50. Alex Prichard and Owen Worth, "Left-Wing Convergence: An Introduction," *Capital and Class* 40, no. 1 (2016): 10.

51. Chris McGreal, "'The S-Word': How Young Americans Fell in Love with Socialism," *Guardian*, September 2, 2017, https://www.theguardian.com/us-news/2017/sep/02/socialism-young-americans-bernie-sanders.

52. Alexander Burns and Maggie Haberman, "Working Families Party Endorses Bernie Sanders for President," *First Draft*, *New York Times*, December 8, 2015, https://www.nytimes.com/politics/first-draft/2015/12/08/working-families-party-endorses-bernie-sanders-for-president/. For more information, see the Working Family Party's website, http://workingfamilies.org/. For an account of Demcoratic Party reformists' realizations, see Asawin Suebsaeng, "Team Bernie on Trump: We Told You So," *Daily Beast*, November 9, 2016, http://www.thedailybeast.com/articles/2016/11/09/team-bernie-on-trump-we-told-you-so.html. And Karen Bernal, et al., *Autopsy: The Democratic Party in Crisis* (San Geronimo, CA: Action for a Progressive Future, 2017), https://democraticautopsy.org/.

53. Notably through the magazine *Jacobin* and in other left media such as, *n+1*, *Triple Canopy*, *The New Inquiry*, *Lies*, and *Lana Turner*. For an overview, see Susan Watkins, "Oppositions," *New Left Review*, no. 98 (2016): 5–30.

See also Reed, "Marx, Race, and Neoliberalism"; Reed, "The Black-Labor-Left Alliance in the Neoliberal Age," *New Labor Forum* 25, no. 2 (2016): 28–34; Ryan Conrad, ed., *Against Equality: Queer Critiques of Gay Marriage* (Chicago: Against Equality Collective, 2015); Michaels, *The Trouble with Diversity*; Yasmin Nair, "The Postracial Delusion," *Monthly Review*, February 2016. Žižek, *In Defense of Lost Causes*; Jodi Dean, *The Communist Horizon* (London: Verso, 2012); Dean, *Crowds and Party* (London: Verso, 2016); and Sean Sweeney, "Contested Futures: Labor after Keystone XL," *New Labor Forum* 25, no. 2 (2016): 93–97.

54. Dixon, *Another Politics*, 226–27 (emphasis in original).

55. Ibid., 228 (emphasis in original).

56. Cornell, *Oppose and Propose!* 162, 179.

57. Ibid., 155–56, 177.

58. Ibid., 177 (emphasis in original).

59. See Lisa Duggan, *The Twilight of Equality? Neoliberalism, Cultural Politics, and the Attack on Democracy* (Boston: Beacon, 2003); Julian Agyeman, *Sustainable Communities and the Challenge of Environmental Justice* (New York: New York University Press, 2005); Laura Pulido, *Environmentalism and Economic Justice: Two Chicano Struggles in the Southwest* (Tuscon: University of Arizona Press, 1996); Kristin Shrader-Frechette, *Environmental Justice: Creating Equality, Reclaiming Democracy* (Oxford: Oxford Univeristy Press, 2002); David Schlosberg, *Environmental Justice and the New Pluralism: The Challenge of Difference for Environmentalism* (Oxford: Oxford University Press, 1999); Schlosberg, *Defining Environmental Justice: Theories, Movements, and Nature* (Oxford: Oxford University Press, 2007); Robert D. Bullard, ed., *Growing Smarter: Achieving Livable Communities, Environmental Justice, and Regional Equity* (Cambridge, MA: MIT Press, 2007); Dorceta E. Taylor, "The Rise of the Environmental Justice

Paradigm: Injustice Framing and the Social Construction of Environmental Discourses," *American Behavioral Scientist* 43, no. 4 (2000): 508–80; and Peter S. Wenz, *Environmental Justice* (Albany: State University of New York Press, 1988).

60. Erik Swyngedouw and Nikolas C. Heynen, "Urban Political Ecology, Justice and the Politics of Scale," *Antipode* 35, no. 5 (2003): 910; Holifield, "Neoliberalism and Environmental Justice Policy," 203 (emphases in original).

61. Pulido, Kohl, and Cotton, "State Regulation and Environmental Justice," 1, 15, 16 (emphases in original).

62. Ibid., 6.

63. Angela Park, *Everybody's Movement: Environmental Justice and Climate Change* (Washington, DC: Environmental Support Center, 2009), 4.

64. David Schlosberg and Lisette Collins, "From Environmental to Climate Justice: Climate Change and the Discourse of Environmental Justice," *WIREs Climate Change* 5, no. 3 (2014): 364.

65. James DeFilippis, Robert Fisher, and Eric Shragge, *Contesting Community: The Limits and Potential of Local Organizing* (New Brunswick, NJ: Rutgers University Press, 2010), 67, 99ff. In the same year that DeFilippis, Fisher, and Shragge's book was published, ACORN was defunded by the fedearl government in response to the activities of rightwing activists. See, http://www.rightwingwatch.org/post/right-wing-acorn-activist-arrested/.

66. Ibid., 20.

67. Ibid., 134.

68. Ibid., 119, 29. These are trends that perhaps began in the 1970s with the consolidation of the New Right in the wake of the crisis of democracy narratives and took off in the 1980s with the "astroturfing" Wise Use movements. See my discussion of the Wise Use movement in the context of the Reagan Revolution in chapter 3. On the effectiveness of rightwing organizing, see Skocpol, *Diminished Democracy*; and Theda Skocpol and Vanessa Williamson, *The Tea Party and the Remaking of Republican Conservatism* (Oxford: Oxford University Press, 2012).

69. Ibid., 122, 132.

70. DeFilippis, Fisher, and Shragge, *Contesting Community*, 164.

71. See Robert Fisher, et al., "'We Are Radical': The Right to the City Alliance and the Future of Community Organizing," *Journal of Sociology and Social Welfare* 40, no. 1 (2013): 157–82. Harmony Goldberg helpfully explains,

> The Right to the City Alliance [was] initiated by the Miami Workers' Center (MWC), Strategic Action for a Just Economy (SAJE) and Tenants and Workers United (TWU) in 2007, the Right to the City Alliance brought together organizations from across the country that were organizing against gentrification in working class

> communities of color. RTTC member organizations include organizations from Boston (ACE, City Life/Vida Urbana, and the Chinese Progressive Association), Los Angeles (Collective Space, East Los Angeles Housing Corporation, Esperanza Community Housing Corporation, Little Tokyo Service Center, Los Angeles Community Action Network, Pilipino Workers Center, SAJE/Strategic Action for A Just Economy, South Asian Network and Union de Vecinos), Miami (Miami Workers Center and Power U Center), New Orleans (Safe Streets), New York (CAAAV/Organizing Asian Communities, Community Voices Heard, FIERCE, FUREE/Families United for Racial and Economic Equality, Jews for Racial and Economic Justice, Make the Road NY, Mothers on the Move, Picture the Homeless, St. Nick's CDC, and WEACT for Environmental Justice), Oakland (Just Cause Oakland), Providence (DARE/Direct Action for Rights and Equality, and the Olneyville Neighborhood Association, San Francisco (POWER/People Organized to Win Employment Rights, PODER, St. Peter's Housing Committee) and the Washington DC metropolitan area (ONE DC and Tenants and Workers United). Right to the City also engages researchers, academics, lawyers and allies to support the work of the base-building organizations.

See Harmony Goldberg, "Building Power in the City: Reflections on the Emergence of the Right to the City Alliance and the National Domestic Worker's Alliance," in *In the Middle of a Whirlwind: 2008 Convention Protests, Movement and Movements*, https://inthemiddleofthewhirlwind.wordpress.com/building-power-in-the-city/.

72. DeFilippis, Fisher, and Shragge, *Contesting Community*, 154–55. In this respect, turf wars between such groups are of course undesirable but inevitable.

73. Jane McAlevey, "The Crisis of New Labor and Alinsky's Legacy: Revisiting the Role of the Organic Grassroots Leaders in Building Powerful Organizations and Movements," *Politics and Society* 43, no. 3 (2015): 415–41.

74. Steve Meacham, "City Life and Occupy: A Developing Relationship," *Shelterforce*, April 24, 2012, http://www.shelterforce.org/article/2644/city_life_and_occupy_a_developing_relationship/.

75. As explained in Nobody Leaves Mid-Hudson's webpage, https://nobodyleavesmidhudson.org/our-mission-and-history/.

76. Nobody Leaves Mid-Hudson, "Just Utilities: Organizing for Solutions to the Household Energy Crisis," executive summary (Poughkeepsie, NY: NLMH, 2016), 3.

77. At the time of writing, member organizations of the New York Energy Democracy Alliance included

Affordable Housing Partnership Homeownership Center; Alliance for a Green Economy (AGREE); Alliance for a Greater New York (ALIGN); Binghamton Regional Sustainability Coalition; Bronx Cooperative Development Initiative/Emerald Cities Collaborative New York City; Catskill Mountainkeeper; Center for Social Inclusion (CSI); Citizens Environmental Coalition; Citizen Action of New York; Citizens for Local Power*; Community Voices Heard; DE-Squared; Good Old Lower East Side (GOLES); Hudson River Sloop Clearwater; Long Island Progressive Coalition; New York State Sustainable Business Council; Nobody Leaves Mid-Hudson; Northwest Bronx Community Clergy Coalition (NWBCCC); People United for Sustainable Housing (PUSH – Buffalo); Sustainable South Bronx; Public Policy & Education Fund of New York (PPEF); Sane Energy Project; Solar One; Syracuse United Neighbors (SUN); Voices of Community Activists and Leaders (VOCAL); Vote Solar.

These links from the organization's home page, http://energydemocracyny.org/members.

78. See Nobody Leaves Mid-Hudson, *Just Utilities; Organizing for Solutions to the Household Energy Crisis*, full report (Poughkeepsie, NY: NLMH, 2016), 50.

79. Ibid., 47.

Conclusion

1. Of course at the time of writing the extent to which so-called alt-Right scholars and activists have wittingly or unwittingly depended on foreign, specifically Russian, support in addition to well-documented domestic support remains an open question; Jane Mayer has researched domestic support for righ-twing groups. See Mayer, *Dark Money*; Mike Lofgren alleges that Russia supports rightwing groups in the United States. See, Mike Lofgren, "Trump, Putin and the Alt-Right International," *The Atlantic*, October 31, 2016, https://www.theatlantic.com/international/archive/2016/10/trump-putin-alt-right-comintern/506015/.

2. Williams, *Beginning*, 12.

3. Williams and Geuss both suggest that early critical theory focused too intently on the theory of freedom. See Williams, *Ethics and the Limits of Philosophy*, 166, 220n; and Geuss, *World without Why*, 103. On freedom as a ration concept that is tied to concepts of justice and legitimation, see Williams, *Beginning*, 80–82.

4. William A. Galston lauds Mouffe as "a leading arch-realist." See Galston, "Realism in Political Theory," *European Journal of Political Theory* 9, no. 4 (2010): 396.

5. As elaborated in, Ernesto Laclau and Chantal Mouffe, *Hegemony and Socialist Strategy: Towards a Radical Democratic Politics* (London: Verso, 1985); Mouffe, *The Return of the Political*; Mouffe, *The Challenge of Carl Schmitt* (London: Verso, 1999); Mouffe, *The Democratic Paradox*; and Mouffe, *On the Political*. On the relationship between Heidegger and Schmitt, see Martin Heidegger and Carl Schmitt, "Heidegger and Schmitt: The Bottom Line," *Telos* 72 (1987): 132.

6. Mouffe, *The Return of the Political*, 17, 74ff. Second quote in paragraph is from Mouffe, "The Importance of Engaging the State," in *What Is Radical Politics Today?* edited by Jonathan Pugh (Basingstoke, UK: Palgrave Macmillan, 2009), 235.

7. Mouffe, "The Importance of Engaging the State," 236–37.

8. As well as Mouffe and Brown, McNay includes in her engagement with agonistic theories work by Linda Zerilli, Jacques Rancière, William Connolly, and James Tully. See Lois McNay, *The Misguided Search for the Political: Social Weightlessness in Radical Democratic Theory* (Cambridge, UK: Polity, 2014).

9. Geuss, *Politics and the Imagination*, 42.

10. Adorno, *Negative Dialectics*, 17–18.

11. The Hobbesian distinction is elaborated most clearly by Quentin Skinner. See Skinner, *Hobbes and Republican Liberty* (Cambridge: Cambridge University Press, 2008), 126.

Bibliography

AccountAbility. *Accountability—Trends in Corporate Responsibility Reporting*. Institute of Social and Ethical Accountability, http://www.accountability.org/advisory-services/strategy-and-governance/.

Adger, W. Neil, and Andrew Jordan. "Sustainability: Exploring the Processes and Outcomes of Governance." In *Governing Sustainabillity*, edited by W. Neil Adger and Andrew Jordan, 3–31. Cambridge: Cambridge University Press, 2009.

Adorno, Theodor W. *Minima Moralia: Reflections from Damaged Life*. Translated by E. F. N. Jephcott. London: Verso, 1974.

———. *Negative Dialectics*. Translated by E. B. Ashton. London: Routledge & Kegan Paul Ltd., [1966] 1973.

———. "Resignation." *Telos*, no. 35 (1978): 165–68.

———. "Progress," in *Can One Live after Auschwitz: A Philosophical Reader*. Edited by Rolf Tiedemann. Stanford, CA: Stanford University Press, [1964] 2003, 126–45.

———, and Herbert Marcuse. "Correspondence on the German Student Movement." *New Left Review* 233 ([1968–69]1999): 123–36.

Agrawal, Arun. *Environmentality: Technologies of Government and the Making of Subjects*. Durham, NC: Duke University Press, 2005.

Agyeman, Julian. *Sustainable Communities and the Challenge of Environmental Justice*. New York: New York University Press, 2005.

Alexander, Michelle. "Black Lives Shattered." *Nation*, February 29, 2016.

———. *The New Jim Crow: Mass Incarceration in the Age of Colorblindness*. New York: New Press, 2010.

Alford, C. Fred. *Science and the Revenge of Nature: Marcuse and Habermas*. Gainesville, FL: University Presses of Florida, 1985.

Althusser, Louis, Étienne Balibar, Roger Establet, Jacques Rancière, and Pierre Macherey. *Reading Capital: The Complete Edition*. Translated by Ben Brewster and David Fernbach (London: Verso, 2016).

Amadae, S. M. *Rationalizing Capitalist Democracy: The Cold War Origins of Rational Choice Liberalism*. Chicago: University of Chicago Press, 2003.

———. *Prisoners of Reason: Game Theory and Neoliberal Political Economy*. New York: Cambridge University Press, 2016.

Atkins, Curtis. "The Third Way International." *Jacobin*, February 11, 2016.

Bakan, Joel. *The Corporation: The Pathological Pursuit of Profit and Power*. Toronto: Viking Canada, 2004.

Banerjee, Subhabrata Bobby. *Corporate Social Responsibility: The Good, the Bad and the Ugly*. Northampton, MA: Edward Elgar, 2007.

Barry, John. "Ecological Modernisation." In *Debating the Earth: The Environmental Politics Reader*, 2nd ed., edited by John S. Drysek and David Schlosberg, 303–22. Oxford: Oxford University Press, 2005.

Bateson, Gregory. *Steps to an Ecology of Mind*. Frogmore, UK: Paladin, 1973.

Beck, Ulrich. *Risk Society: Towards a New Modernity*. Translated by Mark Ritter. London: Sage, 1992.

Beck, Ulrich, and Elisabeth Beck-Gernsheim. *Individualization: Institutionalized Individualism and Its Social and Political Consequences*. Translated by Patrick Camiller. London: Sage, 2002.

Benioff, Marc, and Karen Southwick. *Compassionate Capitalism: How Corporations Can Make Doing Good an Integral Part of Doing Well*. Franklin Lakes, NJ: Career Press, 2003.

Bennett, Jane. *Vibrant Matter: A Political Ecology of Things*. Durham, NC: Duke University Press, 2010.

Berman, Marshall. *All That Is Solid Melts Into Air: The Experience of Modernity*. New York: Simon & Schuster, 1982.

Bernal, Karen, Pia Gallegos, Sam McCann, and Norman Solomon. *Autopsy: The Democratic Party in Crisis*. San Geronimo, CA: Action for a Progressive Future, 2017.

Bey, Hakim. *T.A.Z.: The Temporary Autonomous Zone, Ontological Anarchy, Poetic Terrorism*. New York: Autonomedia, 1991.

Biro, Andrew. *Denaturalizing Ecological Politics: Alienation from Nature from Rousseau to the Frankfurt School and Beyond*. Toronto: University of Toronto Press, 2005.

———. "The Good Life in the Greenhouse? Autonomy, Democracy, and Citizenship in the Anthropocene." *Telos*, no. 172 (2015): 15–37.

Block, Fred, and Margaret R. Somers. *The Power of Market Fundamentalism: Karl Polanyi's Critique*. Cambridge, MA: Harvard University Press, 2014.

Blühdorn, Ingolfur. "Beyond Criticism and Crisis: On the Post-critical Challenge of Niklas Luhmann." *Debatte* 7, no. 2 (1999): 185–99.

———. *Post-ecologist Politics: Social Theory and the Abdication of the Ecologist Paradigm*. London: Routledge, 2000.

Blühdorn, Ingolfur, and Ian Welsh. "Eco-politics Beyond the Paradigm of Sustainability: A Conceptual Framework and Research Agenda." *Environmental Politics* 16, no. 2 (2007): 185–205.

Bobbio, Norberto. *Left and Right: The Significance of a Political Distinction*. Translated by Allan Cameron. Cambridge, UK: Polity, 1996.

Boggs, Carl. "Marxism, Prefigurative Communism, and the Problem of Workers' Control." *Radical America* 11, no. 6 (1977): 99–122.

Boltanski, Luc, and Eve Chiapello. *The New Spirit of Capitalism*. Translated by Gregory Elliott. London: Verso, [1999] 2005.

Boltanski, Luc, and Laurent Thévenot. *On Justification: Economies of Worth*. Translated by Catherine Porter. Princeton, NJ: Princeton University Press, [1991] 2006.

Bookchin, Murray. "Beyond Neo-Marxism." *Telos*, no. 36 (1978): 5–28.

———. "The Communalist Project." *Harbinger* 3, no. 1 (2002): 1–13.

———. "The Concept of Social Ecology." *Coevolution Quarterly* (Winter 1981): 15–22.

———. *The Ecology of Freedom: The Emergence and Dissolution of Hierarchy*. Oakland, CA: AK Press, [1982] 2005.

———. *The Next Revolution: Popular Assemblies and the Promise of Direct Democracy*. Edited by Debbie Bookchin and Blair Taylor. London: Verso, 2015.

———. *Post-scarcity Anarchism*. Montreal: Black Rose Books, 1986.

Bourke, Richard, and Raymond Geuss, eds. *Political Judgement: Essays for John Dunn*. Cambridge: Cambridge University Press, 2009.

Bowen, Howard R. *The Social Responsibilities of the Businessman*. New York: Harper and Row, 1953.

Breines, Wini. *Community and Organization in the New Left, 1962–1968: The Great Refusal*, 2nd ed. New Brunswick, NJ: Rutgers University Press, 1989.

Brenner, Neil. *New State Spaces: Urban Governance and the Rescaling of Statehood*. Oxford: Oxford University Press, 2004.

Brenner, Neil, and Nik Theodore, eds. *Spaces of Neoliberalism: Urban Restructuring in North America and Western Europe*. London: Wiley-Blackwell, 2003.

Brenner, Robert. *The Boom and the Bubble: The US in the World Economy*. London: Verso, 2005.

Brown, Wendy. *Undoing the Demos: Neoliberalism's Stealth Revolution*. New York: Zone Books, 2015.

Brulle, Robert J. *Agency, Democracy, and Nature: The U.S. Environmental Movement from a Critical Theory Perspective*. Cambridge, MA: MIT Press, 2000.

Brunkhorst, H. *Adorno and Critical Theory*. Cardiff: University of Wales Press, 1999.

Brunkhorst, Hauke. *Critical Theory of Legal Revolutions: Evolutionary Perspectives*. London: Bloomsbury, 2014.

Bullard, Robert D., ed. *Growing Smarter: Achieving Livable Communities, Environmental Justice, and Regional Equity*. Cambridge, MA: MIT Press, 2007.

Caldwell, Bruce. Introduction to *The Road to Serfdom: Text and Documents*, by Friedrich Hayek. Definitive ed., edited by Bruce Caldwell. Chicago: University of Chicago Press, 2007.

Callicott, J. Baird. *Beyond the Land Ethic: More Essays in Environmental Philosophy*. Albany: State University of New York Press, 1999.

———. *In Defense of the Land Ethic: Essays in Environmental Philosophy*. Albany: State University of New York Press, 1989.

———. "Intrinsic Values, Quantum Theory, and Environmental Ethics." *Environmental Ethics* 7, no. 3 (1985): 257–75.

———. "The Metaphysical Implications of Ecology." *Environmental Ethics* 8, no. 4 (1986): 301–16.

Campbell, C. *The Romantic Ethic and the Spirit of Consumerism*. Oxford: Blackwell, 1987.

Capra, Frjtjof. *The Tao of Physics: An Exploration of the Parallels between Modern Physics and Eastern Mysticism*. Berkeley, CA: Shambhala, 1975.

———. *The Turning Point: Science, Society, and the Rising Culture*. New York: Bantam, 1983.

———, and Pier Luigi Luisi. *The Systems View of Life: A Unifying Vision*. Cambridge: Cambridge University Press, 2014.

Carroll, William K., and Jean Philippe Sapinski. "The Global Corporate Elite and the Transnational Policy-Planning Network, 1996–2006: A Structural Analysis." *International Sociology* 25, no. 4 (2010): 501–38.

Cavanagh, Sheila M., Robert W. Hahn, and Robert N. Stavins. *National Environmental Policy During the Clinton Years*. Washington, DC: Resources for the Future, 2001.

Chaves, Elisabeth K. *Reviewing Political Criticism: Journals, Intellectuals, and the State*. Burlington, VT: Ashgate, 2015.

Chesters, Graeme, and Ian Welsh. *Complexity and Social Movements: Multitudes at the Edge of Chaos*. London: Routledge, 2006.

Chiapello, Eve. "Capitalism and Its Criticisms." In *New Spirits of Capitalism? Crises, Justifications, and Dynamics*, edited by Paul du Gay and Glenn Morgan, 60–81. Oxford: Oxford University Press, 2013.

Christoff, Peter. "Ecological Modernisation, Ecological Modernities." *Environmental Politics* 5, no. 3 (1996): 476–500.

———. "Out of Chaos, a Shining Star? Toward a Typology of Green States." In *The State and the Global Ecological Crisis*, edited by John Barry and Robyn Eckersley, 25–52. Cambridge, MA: MIT Press, 2005.

Churchill, Ward, and Jim Vander Wall. *The Cointelpro Papers: Documents from the FBI's Secret Wars against Dissent in the United States*. Boston: South End Press, 1990.

Ciepley, David. "Beyond Public and Private: Toward a Political Theory of the Corporation." *American Political Science Review* 107, no. 1 (2013): 139–58.

Commoner, Barry. *Closing the Circle: Nature, Man, and Technology*. New York: Knopf, 1971.

Conrad, Ryan, ed. *Against Equality: Queer Critiques of Gay Marriage*. Chico, CA: AK Press, 2015.

Constant, Benjamin. *Political Writings*, edited by Biancamaria Fontana. Cambridge: Cambridge University Press, 1988.

Cornell, Andrew. *Oppose and Propose! Lessons from Movement for a New Society*. Oakland, CA: AK Press, 2011.

Cowie, Jefferson. *Stayin' Alive: The 1970s and the Last Days of the Working Class*. New York: New Press, 2010.

Cramer, Phillip F. *Deep Environmental Politics: The Role of Radical Environmentalism in Crafting American Enviromental Policy*. Westport, CT: Praeger, 1998.

Critchley, Simon. *The Faith of the Faithless: Experiments in Political Theology*. London: Verso, 2012.

———. *Infinitely Demanding: Ethics of Commitment, Politics of Resistance*. London: Verso, 2007.

Crouch, Colin. *Post-democracy*. Cambridge, UK: Polity, 2004.

———. *The Strange Non-death of Neoliberalism*. Cambridge, UK: Polity, 2011.

Crozier, Michael, Samuel P. Huntington, and Joji Watanuki. *The Crisis of Democracy: Report on the Governability of Democracies to the Trilateral Commission*. New York: New York University Press, 1975.

Dardot, Pierre, and Christian Laval. *The New Way of the World: On Neoliberal Society*. Translated by Gregory Elliott. London: Verso, 2014.

Darier, Éric, ed. *Discourses of the Environment*. Malden, MA: Blackwell, 1999.

Dauvergne, Peter, and Genevieve. LeBaron. *Protest Inc.: The Corporatization of Activism*. Cambridge, UK: Polity, 2014.

Dauvergne, Peter, and Jane Lister. *Eco-business: A Big Brand Takeover of Sustainability*. Cambridge, MA: MIT Press, 2013.

Davis, Mike. "The New Right's Road to Power." *New Left Review*, no. 128 (1981): 28–49.

———. "The Political Economy of Late-Imperial America." *New Left Review*, no. 143 (1984): 6–38.

Day, Richard J. F. *Gramsci Is Dead: Anarchist Currents in the Newest Social Movements*. London: Pluto, 2005.

Dean, Jodi. *The Communist Horizon*. London: Verso, 2012.

———. *Crowds and Party*. London: Verso, 2016.

DeFilippis, James, Robert Fisher, and Eric Shragge. *Contesting Community: The Limits and Potential of Local Organizing*. New Brunswick, NJ: Rutgers University Press, 2010.

Deleuze, Gilles. *Difference and Repetition*. Translated by Paul Patton. New York.: Columbia University Press, 1994.

———. *A Thousand Plateaus: Capitalism and Schizophrenia*. Translated by Brian Massumi. Minneapolis: University of Minnesota Press, 1987.

———, and Félix Guattari. *Anti-Oedipus: Capitalism and Schizophrenia*. Translated by Robert Hurley, Mark Seem, and Helen R. Lane. London: Athlone, 1977.

Denby, Charles. "Workers Battle Automation." In *The New Left: A Collection of Essays*, edited by Priscilla Long, 151–71. Boston: Porter Sargent, 1969.

Denis, Andy. "A Century of Methodological Individualism, Part 2: Mises and Hayek." Deptartment of Economics, City University London, 2010.

———. "Two Rhetorical Strategies of *Laissez-Faire*." *Journal of Economic Methodology* 11, no. 3 (2004): 341–57.

———. "Was Hayek a Panglossian Evolutionary Theorist: A Reply to Whitman." *Constitutional Political Economy* 13, no. 3 (2002): 275–85.

Devall, Bill, and George Sessions. *Deep Ecology: Living as if Nature Mattered*. Salt Lake City, UT: Peregrine Smith, 1985.

de Waal, Frans. *The Age of Empathy: Nature's Lessons for a Kinder Society*. New York: Harmony Books, 2009.

Dixon, Chris. *Another Politics: Talking across Today's Transformative Movements*. Berkeley, CA: University of California Press, 2014.

diZerega, Gus. "Deep Ecology and Liberalism: The Greener Implications of Evolutionary Liberal Theory." *Review of Politics* 58, no. 4 (1996): 699–734.

Dryzek, John S. *Deliberative Democracy and Beyond: Liberals, Critics, Contestations*. Oxford: Oxford University Press, 2000.

———. *Deliberative Global Politics: Discourse and Democracy in a Divided World*. Cambridge, UK: Polity, 2006.

Dryzek, John S., David Downs, Christian Hunold, and D. Schlosberg, with Hans-Kristian Hernes. *Green States and Social Movements: Environmentalism in the United States, United Kingdom, Germany, and Norway*. Oxford: Oxford University Press, 2003.

du Gay, Paul. *Consumption and Identity at Work*. London: Sage, 1996.

Duggan, Lisa. *The Twilight of Equality? Neoliberalism, Cultural Politics, and the Attack on Democracy*. Boston: Beacon, 2003.

Dumont, Louis. *Essays on Individualism: Modern Ideology in Anthropological Perspective*. Chicago: University of Chicago Press, 1986.

———. *Homo Hierarchicus: The Caste System and Its Implications*. Translated by Mark Sainsbury. London: Paladin, 1972.

Eckersley, Robyn. "Deliberative Democracy, Ecological Representation and Risk: Towards a Democracy of the Affected." In *Democratic Innovation: Deliberation, Representation and Association*, edited by Michael Saward, 117–32. London: Routledge, 2000.

———. *Environmentalism and Political Theory: Toward an Ecocentric Approach*. Albany: State University of New York Press, 1992.

Ehrlich, Paul R. *The Population Bomb*. New York: Ballantine, 1968.

Enzensberger, Hans Magnus. "A Critique of Political Ecology." *New Left Review*, no. 84 (1974): 3–31.

Epstein, Barbara. *Political Protest and Cultural Revolution: Nonviolent Direct Action in the 1970s and 1980s*. Berkeley: University of California Press, 1991.

Faricy, Christopher G. *Welfare for the Wealthy: Parties, Social Spending, and Inequality in the United States*. Cambridge: Cambridge University Press, 2015.

Farley, Robert. "Bill Clinton and the 1994 Crime Bill." *The Wire* (blog). FactCheck.org, a Project of the Annenburg Public Policy Center, April 12, 2016. http://www.factcheck.org/2016/04/bill-clinton-and-the-1994-crime-bill/.

Feaver, Peter. "Now I Remember Why President Bush Urged People to Go About Their Daily Lives." *Foreign Policy*, April 17, 2013.

Fisher, Dana R., and William R. Freudenberg. "Ecological Modernisation and Its Critics: Assessing the Past and Looking towards the Future." *Society and Natural Resources* 14, no. 8 (2001): 701–9.

Fisher, Robert, Yuseph Katiya, Christopher Reid, and Eric Shragge. "'We Are Radical': The Right to the City Alliance and the Future of Community Organizing." *Journal of Sociology and Social Welfare* 40, no. 1 (2013): 157–82.

Foucault, Michel. *Discipline and Punish: The Birth of the Prison*. Translated by Alan M. Sheridan. London: Penguin, 1991.

———. *The History of Sexuality*, vol. 3, *The Care of the Self*. Translated by Robert Hurley. London: Penguin, 1986.

———. *The Birth of Biopolitics*. Translated by Graham Burchell. New York: Picador, 2008.

Frank, Thomas. *The Conquest of Cool: Business Culture, Counterculture, and the Rise of Hip Consumerism*. Chicago: University of Chicago Press, 1997.

———. *Listen, Liberal or, What Ever Happened to the Party of the People?* New York: Metropolitan Books, 2016.

———. *One Market under God: Extreme Capitalism, Market Populism, and the End of Economic Democracy*. London: Vintage, 2000.

———. *Pity the Billionaire: The Hard-Times Swindle and the Unlikely Comeback of the Right*. New York: Picador, 2012.

———. *The Wrecking Crew: How Conservatives Rule*. New York: Henry Holt, 2008.

Frankel, Boris. *The Post-industrial Utopians*. Cambridge, UK: Polity, 1987.

Fraser, Nancy. *Unruly Practices: Power, Discourse, and Gender in Contemporary Social Theory*. Cambridge, UK: Polity, 1989.

———. "Rethinking the Public Sphere: A Contribution to the Critique of Actually Existing Democracy." In *Habermas and the Public Sphere*, edited by Craig Calhoun, 109–42. Cambridge, MA: MIT Press, 1992.

———. *Fortunes of Feminism: From State-Managed Capitalism to Neoliberal Crisis*. London: Verso, 2013.

———. "Legitimation Crisis? On the Political Contradictions of Financialized Capitalism." *Critical Historical Studies* 2, no. 2 (2015): 157–89.

Fraser, Steve. *The Age of Acquiescence: The Life and Death of American Resistance to Organized Wealth and Power*. New York: Little, Brown and Company, 2015.

———. *The Limousine Liberal: How an Incendiary Image United the Right and Fractured America*. New York: Basic Books, 2016.

Fraser, Steve, and Gary Gerstle, eds. *The Rise and Fall of the New Deal Order, 1930–1980* Princeton, NJ: Princeton University Press, 1989.

Friedman, Milton. "The Social Responsibility of Business Is to Increase Profits." *New York Times Magazine*, September 13, 1970.

Galbraith, John Kenneth. *Money: Whence It Came, Where It Went*. Boston: Houghton Mifflin, 1975.

Galston, William A. "Realism in Political Theory." *European Journal of Political Theory* 9, no. 4 (2010): 385–411.

Gare, Arran. "From Kant to Schelling to Process Metaphysics: On the Way to Ecological Civilization." *Cosmos and History* 7, no. 2 (2011): 26–69.

———. *The Philosophical Foundations of Ecological Civiliation: A Manifesto for the Future*. London: Routledge, 2016.

Geismer, Lily. "Atari Democrats." *Jacobin*, February 8, 2016.

———. *Don't Blame Us: Suburban Liberals and the Transformation of the Democratic Party*. Princeton, NJ: Princeton University Press, 2015.

Geuss, Raymond. *History and Illusion in Politics*. Cambridge: Cambridge University Press, 2001.

———. *The Idea of a Critical Theory: Habermas and the Frankfurt School*. Cambridge: Cambridge University Press, 1981.

———. *Morality, Culture, and History: Essays on German Philosophy*. Cambridge: Cambridge University Press, 1999.

———. *Outside Ethics*. Princeton, NJ: Princeton University Press, 2005.

———. "Philosophical Anthropology and Social Criticism." *Reification: A New Look at an Old Idea*, edited by Martin Jay, 120–31. Oxford: Oxford University Press, 2008.

———. *Philosophy and Real Politics*. Princeton, NJ: Princeton University Press, 2008.

———. *Politics and the Imagination*. Princeton, NJ: Princeton University Press, 2010.

———. *Public Goods, Private Goods*. With a new preface by the author. Princeton, NJ: Princeton University Press, 2001.

———. *Reality and Its Dreams*. Cambridge, MA: Harvard University Press, 2016.

———. *A World without Why*. Princeton, NJ: Princeton University Press, 2014.

Gibson-Graham, J. K. *The End of Capitalism (as We Knew It): A Feminist Critique of Political Economy*. Oxford: Blackwell, 2006.

———. *A Postcapitalist Politics*. Minneapolis: University of Minnesota Press, 2006.

Giddens, Anthony. *The Class Structure of the Advanced Societies*. London: Hutchinson & Co. Ltd., 1974.

Gill, Stephen. *American Hegemony and the Trilateral Commission*. Cambridge: Cambridge University Press, 1990.

———. "Market Civilization, New Constitutionalism and World Order." In *New Constitutionalism and World Order*, edited by Stephen Gill and A. Claire Cutler, 29–44. Cambridge: Cambridge University Press, 2015.

———. "New Constitutionalism, Democratisation and Global Political Economy." *Pacifica Review* 10, no. 1 (1998): 23–38.

Goldberg, Harmony. "Building Power in the City: Reflections on the Emergence of the Right to the City Alliance and the National Domestic Worker's Alliance." *In the Middle of a Whirlwind: 2008 Convention Protests, Movement and Movements*. https://inthemiddleofthewhirlwind.wordpress.com/building-power-in-the-city/.

Goodin, Robert E. *Innovating Democracy: Democratic Theory and Practice after the Deliberative Turn*. Oxford: Oxford University Press, 2008.

Goodin, Robert E., and John S. Dryzek. "Deliberative Impacts: The Macro-political Uptake of Mini-publics." *Politics and Society* 34, no. 2 (2006): 219–44.

Gore, Al. *Earth in the Balance: Ecology and the Human Spirit*. New York: Houghton Mifflin, 1992.

Gottfried, Paul, and Thomas Fleming. *The Conservative Movement*. Boston: Twayne, 1988.

Gottlieb, Robert. *Forcing the Spring: The Transformation of the American Environmental Movement*, 2nd ed. Washington, DC: Island Press, 2005.

Graeber, David. *The Democracy Project: A History, a Crisis, a Movement*. New York: Spiegel and Grau, 2013.

———. *Fragments of an Anarchist Anthropology*. Chicago: Prickly Paradigm Press, 2004.

———. "The New Anarchists." *New Left Review*, no. 13 (2002): 61–73.

GRI. "Dutch Decade 2: In Conversation with Jan-Peter Balkenende on National Sustainability Day." News and Press Center, GRI website, October 10, 2012. https://www.globalreporting.org/information/news-and-press-center/Pages/Dutch-decade-2-in-conversation-with-Jan-Peter-Balkenende-on-national-sustainability-day.aspx.

GRI, UNEP, KPMG, and UCGA. "Carrots and Sticks—Promoting Transparency and Sustainability: An Update on Trends in Voluntary and Mandatory Approaches to Sustainability Reporting." Nairobi: Global Reporting Initiative/United Nations Environment Programme/KPMG/Unit for Corporate Governance in Africa, 2009.

Grundmann, Reiner. "The Ecological Challenge to Marxism." *New Left Review*, no. 187 (1991): 103–20.

Habermas, Jürgen. *Between Facts and Norms: Contributions to a Discourse Theory of Law and Democracy*. Translated by William Rehg. Cambridge, MA: MIT Press, 1994.

———. *Knowledge and Human Interests*. Translated by Jeremy J. Shapiro. Boston: Beacon, 1971.

———. *Legitimation Crisis*. Translated by Thomas McCarthy. Boston: Beacon, 1975.

———. *Lifeworld and System: A Critique of Functionalist Reason*, vol. 2 of *The Theory of Communicative Action*. Translated by Thomas McCarthy. Boston: Beacon, 1987.

———. "New Social Movements." *Telos*, no. 49 (1981): 33–37.

———. *The Philosophical Discourse of Modernity: Twelve Lectures*. Translated by Frederick G. Lawrence. Cambridge, MA: MIT Press, 1990.

———. *Reason and the Rationalization of Society*, vol. 1 of *The Theory of Communicative Action*. Translated by Thomas McCarthy. Boston: Beacon, 1984.

———. *The Structural Transformation of the Public Sphere: An Inquiry into a Category of Bourgeois Society*. Translated by Thomas Burger and Frederick Lawrence. Cambridge, MA: MIT Press, 1989.

———. *Toward a Rational Society: Student Protest, Science, and Politics*. Translated by Jeremy J. Shapiro. London: Heinemann, 1971.

Hacker, Jacob S., and Paul Pierson. *American Amnesia: How the War on Government Led Us to Forget What Made America Prosper*. New York: Simon and Schuster, 2016.

———. *Winner-Take-All Politics: How Washington Made the Rich Richer—and Turned Its Back on the Middle Class*. New York: Simon and Schuster, 2010.

Hajer, Maarten. "Policy without Polity? Policy Analysis and the Institutional Void." *Policy Sciences* 36, no. 2 (2003): 175–95.

———. *The Politics of Environmental Discourse: Ecological Modernization and the Policy Process*. Oxford: Clarendon Press, 1995.

Hajer, Maarten, and Hendrik Wagenaar, eds. *Deliberative Policy Analysis: Understanding Governance in the Network Society*. Cambridge: Cambridge University Press, 2003.

Hall, Edward. "Bernard Williams and the Basic Legitimation Demand: A Defence." *Political Studies* 63, no. 2 (2015): 466–80.

———. "How to Do Realistic Political Theory (and Why You Might Want To)." *European Journal of Political Theory* 16, no. 3 (2017): 283–303.

Hall, Rodney Bruce, and Thomas J. Biersteker, eds. *The Emergence of Private Authority in Global Governance*. Cambridge: Cambridge University Press, 2002.

Hancock, David. "Neoconservatism, Bohemia and the Moral Economy of Neoliberalism." *Journal for Cultural Research* 20, no. 2 (2016): 101–21.

Hanegraaff, Wouter J. *New Age Religion and Western Culture: Esotericism in the Mirror of Secular Thought.*Albany: State University of New York Press, 1998.

Hardin, Garrett. "Extensions of 'The Tragedy of the Commons.'" *Science*, May 1, 1998, 682–83.

———. "The Tragedy of the Commons." *Science*, December 13, 1968, 1243–48.

Hardt, Michael, and Antonio Negri. *Commonwealth*. Camrbidge, MA: Belknap Press of Harvard University Press, 2009.

———. *Empire*. Cambridge, MA: Harvard University Press, 2000.

———. *Multitude: War and Democracy in the Age of Empire*. London: Penguin, 2005.

Hart, Stuart L. "Beyond Greening: Strategies for a Sustainable World." *Harvard Business Review* 75, no. 1 (1997): 66–76.

Harvey, David. *The Condition of Postmodernity: An Enquiry into the Origins of Cultural Change*. Oxford: Blackwell, 1990.

———. "The Nature of Environment: The Dialectics of Social and Environmental Change." *Socialist Register* 29 (1993): 1–51.

Hawken, Paul. *Next Economy*. New York: Ballantine, 1983.

Hawken, Paul, Amory Lovins, and L. Hunter Lovins. *Natural Capitalism: Creating the Next Industrial Revolution*. Boston: Back Bay Books, 2000.

Hay, Peter. *Main Currents in Western Environmental Thought*. Sydney: University of New South Wales Press, 2002.

Hayek, Friedrich. *The Road to Serfdom*. London: Routledge & Kegan Paul, [1944] 1979.

———. *Studies in Philosophy, Politics, and Economics*. London: Routledge and Kegan Paul, 1967.

———. *The Fatal Conceit: The Errors of Socialism*. Edited by William Warren Bartley. London: Routledge, 1988.

Heelas, Paul. *The New Age Movement: The Celebration of the Self and the Sacralization of Modernity* Oxford: Blackwell, 1996.

Heidegger, Martin. *Being and Time*. Translated by Joan Stambaugh. Albany: State University of New York Press, 1996.

———. "The Question Concerning Technology." Translated by William Lovitt. In *Basic Writings*, edited by David Farrell Krell, 3–35. New York: Harper and Row, 1977.

Heidegger, Martin, and Carl Schmitt. "Heidegger and Schmitt: The Bottom Line." *Telos* 72 (1987): 132.

Heilbroner, Robert L. *An Inquiry into the Human Prospect*. New York: Norton, 1974.

Held, David. *Introduction to Critical Theory: Horkheimer to Habermas*. Berkeley: University of California Press, 1980.

Helvarg, David. *The War against the Greens: The "Wise Up" Movement, the New Right, and the Browning of America*, revised ed. Boulder, CO: Johnson Books, 2004.

Hendriks, Carolyn M. "Deliberative Governance in the Context of Power." *Policy and Society* 28, no. 3 (2009): 173–84.

———. "Institutions of Deliberative Democratic Process and Interest Groups: Roles, Tensions and Incentives." *Australian Journal of Public Administration* 61, no. 1 (2002): 64–75.

———. "When the Forum Meets Interest Politics: Strategic Uses of Public Deliberation." *Politics and Society* 34, no. 4 (2006): 571–602.

Hendriks, Carolyn M., John S. Dryzek, and Christian Hunold. "Turning up the Heat: Partisanship in Deliberative Innovation." *Political Studies* 55, no. 2 (2007): 362–83.

Hilton, Adam. "Searching for a New Politics: The New Politics Movement and the Struggle to Democratize the Democratic Party, 1968–1978." *New Political Science* 38, no. 2 (2016): 141–59.

Hirschman, Albert O. *The Passions and the Interests: Political Arguments for Capitalism before Its Triumph*. 20th anniversary ed. Princeton, NJ: Princeton University Press, 1997.

Hobsbawm, Eric. *The Age of Revolution: Europe 1789–1848*. London: Abacus, 1962.

Holden, Meg, and Andy Scerri. "Justification, Compromise and Test: Developing a Pragmatic Sociology of Critique to Understand the Outcomes of Urban Redevelopment." *Planning Theory* 14, no. 4 (2015): 360–83.

Holden, Meg, Andy Scerri, and Cameron Owens. "More Publics, More Problems: The Productive Interface between the Pragmatic Sociology of Critique and Deweyan Pragmatism." *Contemporary Pragmatism* 10, no. 2 (2013): 1–24.

Holifield, Ryan. "Neoliberalism and Environmental Justice Policy." In *Neoliberal Environments: False Promises and Unnatural Consequences*, edited by Nik Heynen, James McCarthy, Scott Prudham, and Paul Robbins, 202–16. New York: Routledge, 2007.

Holloway, John. *Change the World without Taking Power*. London: Pluto, 2002.

Horkheimer, Max. *Critique of Instrumental Reason: Lectures and Essays since the End of World War II*. New York: Seabury, [1967] 1974.

———. *Eclipse of Reason*. New York: Seabury, 1974.

———. *Critical Theory: Selected Essays*. New York: Continuum, 2002.

Horkheimer, Max, and Theodor W. Adorno. *Dialectic of Enlightenment: Philosophical Fragments*. Translated by Edmund Jephcott. Stanford, CA: Stanford University Press, [1947] 2002.

Hunold, Christian, and John S. Dryzek. "Green Political Strategy and the State: Combining Political Theory and Comparative History." In *The State and the Global Ecological Crisis*, edited by John Barry and Robyn Eckersley, 75–96. Cambridge, MA: MIT Press, 2005.

Hunold, Christian, and John S. Dryzek. "Green Political Theory and the State: Context Is Everything." *Global Environmental Politics* 2, no. 3 (2002): 17–39.

Inglehart, Ronald. *Culture Shift in Advanced Industrial Society*. Princeton, NJ: Princeton University Press, 1990.

———. *The Silent Revolution: Changing Values and Political Styles among Western Publics*. Princeton, NJ: Princeton University Press, 1977.

Innes, Judith E., and David E. Booher. "Collaborative Policymaking: Governance through Dialogue." In *Deliberative Policy Analysis: Understanding Governance in the Network Society*, edited by Maarten Hajer and Hendrik Wagenaar, 33–59. Cambridge: Cambridge University Press, 2003.

The Invisible Committee. *The Coming Insurrection*. Los Angeles: Semiotext(e), 2009.

———. *To Our Friends*. Translated by Robert Hurley. Los Angeles: Semiotext(e), 2014.

James, Paul. *Nation Formation: Towards a Theory of Abstract Community*. London: Sage, 1996.

James, Simon P. "Against Holism." *Environmental Values* 16, no. 4 (2007): 447–61.

Jänicke, Martin. *Ecological Modernization: Innovation and Diffusion of Policy and Technology*. Berlin: Forschungsstelle für Umweltpolitik, Freie Universität Berlin, 2008.

Jay, Martin. *Marxism and Totality: The Adventures of a Concept from Lukács to Habermas*. Berkeley: University of California Press, 1984.

———. *The Dialectical Imagination: A History of the Franfkurt School and the Institute of Social Research, 1923–1950*. Berkeley: University of California Press, 1996.

———. *Reason after Its Eclipse: On Late Critical Theory*. Madison: University of Wisconsin Press, 2016.

Jessop, Bob. "Liberalism, Neoliberalism, and Urban Governance: A State-Theoretical Perspective." *Antipode* 34, no. 3 (2002): 452–72.

Jordan, Andrew, Rüdiger K. W. Wurzel, and Anthony R. Zito, eds. *"New" Instruments of Environmental Governance? National Experiences and Prospects* London: Frank Cass, 2003.

Kant, Immanuel. *An Answer to the Question: What Is Enlightenment? Practical Philosophy*. Translated by Mary J. Gregor. Cambridge, UK: Cambridge University Press, [1784].

Kauffman, L. A. "The Theology of Consensus." *Berkeley Journal of Sociology*, May 26, 2015. http://berkeleyjournal.org/2015/05/the-theology-of-consensus/.

Kaufman, Burton I. *The Presidency of James Earl Carter, Jr.* Lawrence: University Press of Kansas, 1993.

Keucheyan, Razmig. *The Left Hemisphere: Mapping Critical Theory Today*. Translated by Gregory Elliott. London: Verso, 2014.

Kingsnorth, Paul. *One No, Many Yeses: A Journey to the Heart of the Global Resistance Movement*. London: Free Press, 2003.

Klein, Naomi. *No Logo*. New York: Picador, 1999.

———. "It's Hard Work Trying to Escape from the Gooey Corporate Hug." *The Guardian Weekly* 14–20 (international edition), June 2001. https://www.theguardian.com/theguardian/2001/jun/14/guardianweekly.guardianweekly11.

Knorringa, Peter, and A. H. J. Helmsing. "Beyond an Enemy Perception: Unpacking and Engaging the Private Sector." *Development and Change* 39, no. 6 (2008): 1053–1062.

Kolakowski, Leszek. "The Persistence of the Sein-Sollen Dilemma." *Man and World* 10, no. 2 (1977): 194–233.

Kowalsky, Nathan. "Whatever Happened to Deep Ecology?" *Trumpeter* 30, no. 2 (2014): 95–100.

Kristol, Irving. "Ideology and Supply-Side Economics." *Commentary*, 71, no. 4 (1981): 48–54.

———. *Two Cheers for Capitalism*. New York: Basic Books, 1978.

Kuhner, Timothy K. *Capitalism v. Democracy: Money in Politics and the Free Market Constitution*. Stanford, CA: Stanford Law Books, an imprint of Stanford University Press, 2014.

———. "*Citizens United* as Neoliberal Jurisprudence: The Resurgence of Economic Theory." *Virginia Journal of Social Policy and the Law* 18, no. 3 (2011): 395–468.

Kurlansky, Mark. *1968: The Year That Rocked the World*. London: Vintage, 2004.

Laclau, Ernesto, and Chantal Mouffe. *Hegemony and Socialist Strategy: Towards a Radical Democratic Politics*. London: Verso, 1985.

Lafer, G. *The One Percent Solution: How Corporations Are Remaking America One State at a Time*. Ithaca, NY: Cornell University Press/ILR Press, 2017.

Landauer, Gustav. *Revolution, and Other Writings: A Political Reader*. Edited and translated by Gabriel Kuhn. Oakland, CA: PM Press, 2010.

Lefort, Claude. *The Political Forms of Modern Society: Bureaucracy, Democracy, Totalitarianism*. Edited by John B. Thompson. Cambridge, MA: MIT Press, 1986.

Leopold, Aldo. *A Sand County Almanac, with Essays on Conservation from Round River*. New York: Sierra Club, 1966.

Lepori, Matthew. "There Is No Anthropocene: Climate Change, Species-Talk, and Political Economy." *Telos*, no. 172 (2015): 103–24.

Leuchtenburg, William E. "Jimmy Carter and the Post–New Deal Presidency." In *The Carter Presidency: Policy Choices in the Post–New Deal Era*, edited by Gary M. Fink and Hugh Davis Graham, 7–28. Lawrence: University Press of Kansas, 1998.

Lévi-Strauss, Claude. *Tristes Tropiques*. Translated by John Weightmann and Doreen Weightman. London: Penguin, 1976.

Levy, David L., and Daniel Egan. "A Neo-Gramscian Approach to Corporate Political Strategy: Conflict and Accommodation in the Climate Change Negotiations." *Journal of Management Studies* 40, no. 4 (2003): 803–29.

Liévanos, Raoul S. "Certainty, Fairness, and Balance: State Resonance and Environmental Justice Policy Implementation." *Sociological Forum* 27, no. 2 (2012): 481–503.

Light, Andrew, and Holmes Rolston III. "Introduction: Ethics and Environmental Ethics." In *Environmental Ethics: An Anthology*, edited by Andrew Light and Holmes Rolston III, 1–11. Maldon, MA: Blackwell, 2003.

Lilla, Mark. *The Once and Future Liberal: After Identity Politics*. New York: Harper, 2017.

Lloyd, Genevieve. *The Man of Reason: "Male" and "Female" in Western Philosophy*. London: Methuen, 1984.

Locke, J. *An Essay Concerning Human Understanding*. Edited by Peter H. Nidditch for *The Clarendon Edition of the Works of John Locke*. Oxford: Oxford University Press, [1690].

Lofgren, Mike. "Trump, Putin and the Alt-Right International." *The Atlantic*, October 31, 2016. https://www.theatlantic.com/international/archive/2016/10/trump-putin-alt-right-comintern/506015/.

Lordon, Frédéric. *Willing Slaves of Capital: Spinoza and Marx on Desire*. London: Verso, 2014.

Lövbrand, Eva, Johannes Stripple, and Bo Wiman. "Earth System Governmentality: Reflections on Science in the Anthropocene." *Global Environmental Change* 19, no. 1 (2009): 7–13.

Lovelock, James, and Sidney Epton. "The Quest for Gaia." *New Scientist*, February 6, 975, 304–6.

Löwy, Michael, and Robert Sayre. *Romanticism against the Tide of Modernity*. Translated by Catherine Porter. Durham, NC: Duke University Press, 2001.

Luke, Timothy W. "The Anthropocene and Freedom: Terrestrial Time as Political Mystification." *Platypus Review* 60, no. 2 (2013): 2–4.

———. *Capitalism, Democracy, and Ecology: Departing from Marx*. Urbana: University of Illinois Press, 1999.

———. "Corporate Social Responsibility: An Uneasy Merger of Sustainability and Development." *Sustainable Development* 21, no. 2 (2013): 83–91.

———. "Culture and Politics in the Age of Artificial Negativity." *Telos*, no. 35 (1978): 56–72.

———. "Deep Ecology: Living as if Nature Mattered: Devall and Sessions on Defending the Earth." *Organization and Environment* 15, no. 2 (2002): 178–86.

———. "Developing Planetarian Accountancy: Fabricating Nature as Stock, Service, and System for Green Governmentality." In *Nature, Knowledge and Negation*, edited by Harry F. Dahms, 129–59. Bingley, UK: Emerald, 2009.

———. *Ecocritique: Contesting the Politics of Nature, Economy, and Culture*. Minneapolis: University of Minnesota Press, 1997.

———. "Environmentality as Green Governmentality." In *Discourses of the Environment*, edited byÉric Darier, 121–51. Malden, MA: Blackwell, 1999.

———. "Informationalism and Ecology." *Telos*, no. 56 (1983): 59–73.

———. *Social Theory and Modernity: Critique, Dissent, and Revolution*. Thousand Oaks, CA: Sage, 1990.

———. "The System of Sustainable Degradation." *Capitalism, Nature, Socialism* 17, no. 1 (2006): 99–112.

Lukes, Steven. "Epilogue: The Grand Dichotomy of the Twentieth Century." In *The Cambridge History of of Twentieth-Century Political Thought*, edited by Terence Ball and Richard Bellamy, 602–26. Cambridge: Cambridge University Press, 2006.

Lyotard, Jean-François. *The Postmodern Condition: A Report on Knowledge*. Translated by Geoff Bennington and Brian Massumi. Minneapolis: University of Minnesota Press, 1993.

MacIntyre, Alasdair. *After Virtue: A Study in Moral Theory*, 2nd ed. Notre Dame, IN: University of Notre Dame Press, 1986.

———. *Against the Self-Images of the Age: Essays on Ideology and Philosophy*. London: Duckworth, 1971.

———. *Whose Justice? Which Rationality?* London: Duckworth, 1988.

MacLean, Nancy. *Democracy in Chains: The Deep History of the Radical Right's Stealth Plan for America*. New York: Viking, 2017.

Müller, Jan-Werner. *What Is Populism?* Philadelphia: University of Pennsylvania Press, 2016.

Maeckelbergh, Marianne. *The Will of the Many: How the Alterglobalisation Movement Is Changing the Face of Democracy*. London: Pluto, 2011.

Mann, Jim, and Jonathan Peterson. "Anti-War Activists Call Clinton Role Peripheral." *Los Angeles Times*, October 10, 1992. http://articles.latimes.com/1992-10-10/news/mn-689_1_vietnam-war-protest.

Marcuse, Herbert. *Eros and Civilization: A Philosophical Inquiry into Freud*. London: Sphere, 1970.

———. *An Essay on Liberation*. London: Penguin, 1969.

———. *One Dimensional Man*. London: Abacus, 1972.

———. "Repressive Tolerance." In *A Critique of Pure Tolerance*, edited by Robert Paul Wolff, Barrington Moore Jr., and Herbert Marcuse, 93–138. London: Jonathan Cape, 1969.

Mariotti, Shannon. *Adorno and Democracy: the American Years*. Lexington, KY: University Press of Kentucky, 2016.

Mayer, Jane. *Dark Money: The Hidden History of the Billionaries behind the Rise of the Radical Right*. New York: Doubleday, 2016.

McAlevey, Jane. "The Crisis of New Labor and Alinsky's Legacy: Revisiting the Role of the Organic Grassroots Leaders in Building Powerful Organizations and Movements." *Politics and Society* 43, no. 3 (2015): 415–41.

McBarnet, Doreen, Aurora Voiculescu, and Tom Campbell, eds. *The New Corporate Accountability: Corporate Social Responsibility and the Law*. Cambridge: Cambridge University Press, 2007.

McNay, Lois. *The Misguided Search for the Political: Social Weightlessness in Radical Democratic Theory*. Cambridge, UK: Polity, 2014.

Meacham, Jon. *Destiny and Power: The American Odyssey of George Herbert Walker Bush*. New York: Random House, 2015.

Meacham, Steve. "City Life and Occupy: A Developing Relationship." *Shelterforce*, April 24, 2012. http://www.shelterforce.org/article/2644/city_life_and_occupy_a_developing_relationship/.

Meadowcroft, James. "From Welfare State to Ecostate." In *The State and the Global Ecological Crisis*, edited by John Barry and Robyn Eckersley, 3–24. Cambridge, MA: MIT Press, 2005.

Meek, James. "Robin Hood in a Time of Austerity." *London Review of Books*, February 18, 2016.

Menand, Louis. *The Metaphysical Club: A Story of Ideas in America*. New York: Farrar, Straus and Giroux, 2002.

Merchant, Carolyn, ed. *Ecology*. Amherst, NY: Humanity Books, 1999.

Meyer, John M. *Political Nature: Environmentalism and the Interpretation of Western Thought*. Cambridge, MA: MIT Press, 2001.

Meyer, John M., and Jens M. Kersten, eds. *The Greening of Everyday Life: Challenging Practices, Imagining Possibilities*. Oxford: Oxford University Press, 2016.

Michaels, Walter Benn. *The Trouble with Diversity: How We Learned to Love Identity and Ignore Inequality*. New York: Holt, 2006.

Miroff, Bruce. *The Liberals' Moment: The McGovern Insurgency and the Identity Crisis of the Democratic Party*. Lawrence: University Press of Kansas, 2007.

Mirowski, Philip, and Dieter Plehwe, eds. *The Road from Mont Pèlerin: The Making of the Neoliberal Thought Collective*. Cambridge, MA: Harvard University Press, 2009.

Morgan, Edward P. *The Sixties Experience: Hard Lessons about Modern America*. Phiadelphia: Temple University, 1991.

Mouffe, Chantal, ed. *The Challenge of Carl Schmitt*. London: Verso, 1999.

———. *The Democratic Paradox*. London: Verso, 2000.

———. "The Importance of Engaging the State." In *What Is Radical Politics Today?*, edited by Jonathan Pugh, 230–37. Basingstoke, UK: Palgrave Macmillan, 2009.

———. *On the Political*. London: Routledge, 2005.

———. *The Return of the Political*. London: Verso, 1993.

Moyn, Samuel. *The Last Utopia: Human Rights in History*. Cambridge, MA: Belknap Press of Harvard University Press, 2010.

———. "A Powerless Companion: Human Rights in the Age of Neoliberalism." *Law and Contemporary Problems* 77, no. 4 (2014): 147–70.

Mukerji, Chandra. *From Graven Images: Patterns of Modern Materialism*. New York: Columbia University Press, 1983.

Mulholland, Marc. *Bourgeois Liberty and the Politics of Fear: From Absolutism to Neo-conservatism*. New York: Oxford University Press, 2012.

Murakawa, Naomi. *The First Civil Right: How Liberals Built Prison America*. Oxford: Oxford University Press, 2014.

Murphy, Joseph. "Ecological Modernisation." *Geoforum* 31, no. 1 (2000): 1–8.

Musgrove, Frank. *Ecstasy and Holiness: Counter Culture and the Open Society*. London: Methuen, 1974.

Nadasen, Premilla. "How a Democrat Killed Welfare." *Jacobin*, February 9, 2016.

Nadasen, Premilla, Jennifer Mittelstadt, and Marisa Chappell. *Welfare in the United States: A History with Documents, 1935–1996*. New York: Routledge, 2009.
Naess, Arne. *Ecology, Community, and Lifestyle: Outline of an Ecosophy*. Translated by David Rothenberg. Cambridge: Cambridge University Press, 1993.
———. "The Shallow and the Deep, Long-Range Ecology Movement: A Summary." *Inquiry* 16 (1973): 95–100.
Nair, Yasmin. "The Postracial Delusion." *Monthly Review*, February 2016.
Nash, Nicholas, and Alan Lewis. "Overcoming Obstacles to Ecological Citizenship: The Dominant Social Paradigm and Local Environmentalism." In *Environmental Citizenship*, edited by Andrew Dobson and Derek Bell, 153–84. Cambridge, MA: MIT Press, 2006.
Nash, Roderick. "Aldo Leopold's Intellectual Heritage." In *Companion to "A Sand County Almanac": Interpretive and Critical Essays*, edited by J. Baird Callicott, 63–88. Madison: University of Wisconsin Press, 1987.
Nattrass, Brian, and Mary Altomare. *The Natural Step for Business: Wealth, Ecology, and the Evolutionary Corporation*. Gabriola Island, BC: New Society, 1999.
"The New Trade War: Opponents of Globalisation Wreaked Havoc at a Big Trade Summit in Seattle This Week." *Economist*, December 2, 1999. http://www.economist.com/node/264372.
Nickel, Patricia Mooney, ed. *North American Critical Theory after Postmodernism: Contemporary Dialogues*. New York: Palgrave Macmillan, 2012.
Nie, Martin A. "'It's the Environment, Stupid!' Clinton and the Environment." *Presidential Studies Quarterly* 27, no. 1 (1997): 39–51.
Nietzsche, Friedrich. *Beyond Good and Evil: Prelude to a Philosophy of the Future*. Translated by Judith Norman. Cambridge: Cambridge University Press, [1886] 2002.
———. *The Gay Science*. Translated by Josefine Nauckhoff and Adrian Del Caro. Cambridge: Cambridge University Press, [1887] 2001.
Nobody Leaves Mid-Hudson. "Just Utilities: Organizing for Solutuions to the Household Energy Crisis." Executive summary. Poughkeepsie, NY: NLMH, 2016.
———. *Just Utilities: Organizing for Solutuions to the Household Energy Crisis*. Full report. Poughkeepsie, NY: NLMH, 2016.
Nozick, Robert. *Anarchy, State, and Utopia*. New York: Basic Books, 1974.
O'Connor, James. *The Fiscal Crisis of the State*. New York: Transaction, 2009.
Offe, Claus. *Contradictions of the Welfare State*. Cambridge, MA: MIT Press, 1984.
———. "Reflections on the Institutional Self-Transformation of Movement Politics: A Tentative Stage Model." In *Challenging the Political Order: New Social and Political Movements in Democracies*, edited by Russell J. Dalton and Manfred Kuechler, 232–50. Cambridge: Polity, 1990.
———. *Reflections on America: Tocqueville, Weber and Adorno in the United States*. Cambridge: Polity, 2005.

Ophuls, William. *Ecology and the Politics of Scarcity: Prologue to a Political Theory of the Steady State*. San Francisco: W. H. Freeman, 1977.
Paehlke, Robert. "Eco-History: Two Waves in the Evolution of Environmentalism." *Alternatives* 19, no. 1 (1992): 18–23.
Pagden, Anthony. *The Enlightenment: And Why It Still Matters*. New York: Random House, 2013.
Park, Angela. *Everybody's Movement: Environmental Justice and Climate Change*. Washington, DC: Environmental Support Center, 2009.
Parker, Christine. *The Open Corporation: Effective Self-Regulation and Democracy*. Cambridge: Cambridge University Press, 2002.
Parson, Sean. "Parks, Permits, and Riot Police: San Francisco Food Not Bombs and Autonomous Occupations of Space." *New Political Science* 37, no. 3 (2015): 346–62.
Pauli, Benjamin J. "The New Anarchism in Britain and the US: Towards a Richer Understanding of Post-war Anarchist Thought." *Journal of Political Ideologies* 20, no. 2 (2015): 134–55.
Peck, Jamie, and Adam Tickell. "Searching for a New Institutional Fix: The *After*-Fordist Crisis and the Global-Local Disorder." In *Post-Fordism: A Reader*, edited by Ash Amin, 280–316. Oxford: Blackwell, 1994.
Phillips-Fein, Kim. "Conservatism: A State of the Field." *Journal of American History* 98, no. 3 (2011): 723–43.
———. *Invisible Hands: The Making of the Conservative Movement from the New Deal to Reagan*. New York: Norton, 2009.
Piccone, Paul. "The Crisis of One-Dimensionality." *Telos*, no. 35 (1978): 43–54.
Pierre, Jon, ed. *Debating Governance: Authority, Steering, and Democracy*. Oxford: Oxford University Press, 2000.
Pierre, Jon, and B. Guy Peters. *Governance, Politics, and the State*. London: Macmillan, 2000.
Piven, Frances Fox. *Challenging Authority: How Ordinary People Change America*. Lanham, MD: Rowman and Littlefield, 2006.
———, and Richard A. Cloward. *Poor People's Movements: Why They Succeed, How They Fail*. New York: Vintage, 1979.
Plumwood, Val. "Ecosocial Feminism as a General Theory of Social Oppression." In *Ecology: Key Concepts in Critical Theory*, edited by Carolyn Merchant, 207–19. New York: 1999.
———. "Feminism." In *Political Theory and the Ecological Challenge*, edited by Andrew Dobson and Robyn Eckersley, 51–74. Cambridge: Cambridge University Press, 2006.
Pollin, Robert. *Contours of Descent: US Economic Fractures and the Landscape of Global Austerity*, new ed. London: Verso, 2005.
Powell, Lewis F., Jr. "Attack on American Free Enterprise System." Confidential memorandum to Eugene B. Sydnor Jr., August 23, 1971. http://law2.wlu.edu/powellarchives/page.asp?pageid=1251.

Prichard, Alex, and Owen Worth. "Left-Wing Convergence: An Introduction." *Capital and Class* 40, no. 1 (2016): 3–17.

Prinz, Janosch. "Raymond Geuss' Radicalization of Realism in Political Theory." *Philosophy and Social Criticism* 42, no. 8 (2016), 777–96.

Pulido, Laura. *Environmentalism and Economic Justice: Two Chicano Struggles in the Southwest*. Tuscon: University of Arizona Press, 1996.

Pulido, Laura, Ellen Kohl, and Nicole-Marie Cotton. "State Regulation and Environmental Justice: The Need for Strategy Reassessment." *Capitalism, Nature, Socialism* 27, no. 2 (2016): 12–31.

Purdy, Jedediah. *After Nature: A Politics for the Anthropocene*. Cambridge, MA: Harvard University Press, 2015.

Putnam, Robert D. *Bowling Alone: The Collapse and Revival of American Community*. New York: Simon and Schuster, 2000.

Radkau, Joachim. *The Age of Ecology: A Global History*. Translated by Patrick Camiller. Cambridge, UK: Polity, 2014.

———. *Nature and Power: A Global History of the Environment*. Translated by Thomas Dunlap. Cambridge: Cambridge University Press, 2008.

Rancière, Jacques. *Dissensus: On Politics and Aesthetics*. Translated by Steven Corcoran. London: Bloomsbury, 2015.

Rawls, John. *A Theory of Justice*, revised ed. Cambridge, MA: Belknap Press of Harvard University Press, [1971] 2005.

Redclift, Michael. *Sustainable Development: Exploring the Contradictions*. London: Routledge, 1989.

Reed, Adolph, Jr. "The Black-Labor-Left Alliance in the Neoliberal Age." *New Labor Forum* 25, no. 2 (2016): 28–34.

———. "Marx, Race, and Neoliberalism." *New Labor Forum* 22, no. 1 (2013): 49–57.

Reich, Charles A. *The Greening of America*. London: Penguin, 1995.

Reich, Robert B. *Locked in the Cabinet*. New York: Knopf, 1997.

Rhodes, R. A. W. *Understanding Governance: Policy Networks, Governance, Reflexivity, and Accountability*. Buckingham, UK: Open University Press, 1997.

Robin, Corey. *Fear: The History of a Political Idea*. Oxford: Oxford University Press, 2004.

Rodgers, Daniel T. *The Age of Fracture*. Cambridge, MA: Belknap Press of Harvard University Press, 2012.

Roelvink, Gerda, Kevin St. Martin, and J. K. Gibson-Graham, eds. *Making Other Worlds Possible: Performing Diverse Economies*. Minneapolis: University of Minnesota Press, 2015.

Rorty, Richard. *Achieving our Country: Leftist Thought in Twentieth Century America*. Cambridge MA, Harvard University Press, 1998.

Rosenau, James N., and Ernst-Otto. Czempiel, eds. *Governance without Government: Order and Change in World Politics*. Cambridge: Cambridge University Press, 1992.

Rosenbaum, Walter A. *Environmental Politics and Policy*, 9th ed. Washington, DC: CQ Press, 2014.

Ross, Kristin. *May '68 and Its Afterlives*. Chiacgo: University of Chicago Press, 2002.

Rossi, Enzo, and Matt Sleat. "Realism in Normative Political Theory." *Philosophy Compass* 9, no. 10 (2014): 689–701.

Roszak, Theodore. *The Making of a Counterculture: Reflections on the Technocratic Society and Its Youthful Opposition*, with new introduction. Berkeley: University of California Press, 1995.

Rowe, James K. "Corporate Social Responsibility as Business Strategy." In *Globalization, Governmentality and Global Politics: Regulation for the Rest of Us?*, edited by Ronnie D. Lipschutz and James K. Rowe, 130–70. London: Routledge, 2005.

Rubin, Charles T. *The Green Crusade: Rethinking the Roots of Environmentalism*. Lanham, MD: Rowman & Littlefield, 1993.

Runciman, David. *The Confidence Trap: A History of Democracy in Crisis from World War I to the Present*. Princeton, NJ: Princeton University Press, 2013.

Rutherford, Stephanie. "Green Governmentality: Insights and Opportunities in the Study of Nature's Rule." *Progress in Human Geography* 31, no. 3 (2007): 291–307.

Sale, Kirkpatrick. *The Green Revolution: The American Environmental Movement, 1962–1992*. New York: Hill and Wang, 1994.

Sandel, Michael. *Liberalism and the Limits of Justice*, 2nd ed. Cambridge: Cambirdge University Press, 1998.

Sanofi. *Corporate Social Responsibility Report 2012*. Paris: Sanofi, 2012.

Scerri, Andy. "Comfortably Inhabiting Reality: Justifying and Denouncing Arguments in a Development Dispute in the Post-industrial Gentrified Inner-City." *Space and Polity* 18, no. 1 (2014): 39–53.

———. "Deep Ecology, the Holistic Critique of Enlightenment Dualism, and the Irony of History." *Environmental Values* 25, no. 5 (2016): 527–51.

———. "Green Citizenship and the Political Critique of Injustice." *Citizenship Studies* 17, no. 3–4 (2013): 293–307.

———. *Greening Citizenship: Sustainable Development, the State and Ideology*. Basingstoke, UK: Palgrave Macmillan, 2012.

———. "Paradoxes of Increased Individuation and Public Awareness of Environmental Issues." *Environmental Politics* 18, no. 4 (2009): 467–85.

———. "Reaganomics." In *Wiley Encyclopedia of Social Theory*, edited by Brian S. Turner, 1939–41. New York: Wiley, 2017.

———. "Triple Bottom-Line Capitalism and the 3rd Place." *Arena Journal*, no. 20 (2003): 57–66.

———. "The World Social Forum: Another World Might Be Possible." *Social Movement Studies* 12, no. 1 (2013): 111–20.

———, and Cynthia Lam. "From Neoliberalism to Neocommunitarianism: Opposing Justifications in a Dispute over Privatized Electricity Infrastructure." *Space and Polity* 19, no. 2 (2015): 132–49.

———, and Liam Magee. "Green Householders, Stakeholder Citizenship and Sustainability." *Environmental Politics* 21, no. 3 (2012): 387–411.

———, and Nader Sobhani. "Even Natural Disasters Are Unlikely to Slow Us Down: Corporate Social & Environmental Responsibility as Well-Crafted Political Judgment." In *Biopolitical Disaster*, edited by J. Lawrence and S.-M. Wiebe, 62–77. London: Routledge, 2018.

Schaar, John H. "Legitimacy and the Modern State." In *Power and Community: Dissenting Essays in Political Science*, edited by Philip Green and Sanford Levinson, 276–327. New York: Vintage, 1970.

Schelling, Thomas C. "Hockey Helmets, Concealed Weapons and Daylight Saving: A Study of Binary Choices with Externalities." *Journal of Conflict Resolution* 17, no. 3 (1973): 381–428.

Schlosberg, David. *Defining Environmental Justice: Theories, Movements, and Nature*. Oxford: Oxford University Press, 2007.

———. *Environmental Justice and the New Pluralism: The Challenge of Difference for Environmentalism*. Oxford: Oxford University Press, 1999.

———. "Theorising Environmental Justice: The Expanding Sphere of a Discourse." *Environmental Politics* 22, no. 1 (2013): 37–55.

Schlosberg, David, and Romand Coles. "The New Environmentalism of Everyday Life: Sustainability, Material Flows and Movements." *Contemporary Political Theory* 15, no. 2 (2016): 160–81.

Schlosberg, David, and Lisette B. Collins. "From Environmental to Climate Justice: Climate Change and the Discourse of Environmental Justice." *WIREs Climate Change* 5, no. 3 (2014): 359–74.

Schlosberg, David, and John S. Dryzek. "Political Strategies of American Environmentalism: Inclusion and Beyond." *Society and Natural Resources* 15, no. 9 (2002): 787–804.

Schlosberg, David, and Sara Rinfret. "Ecological Modernisation, American Style." *Environmental Politics* 17, no. 2 (2008): 254–75.

Schulman, Bruce J. *The Seventies: The Great Shift in American Culture, Society, and Politics*. New York: Da Capo, 2002.

Scott, Douglas W. "Student Activism on Environmental Crisis." *Living Wilderness* 34 (1970): 231–42.

Selznick, Philip. *The Communitarian Persuasion*. Baltimore: Johns Hopkins University Press, 2002.

Shabecoff, Philip. *A Fierce Green Fire: The American Environmental Movement*. New York: Hill and Wang, 1994.

Shamir, R. "The Age of Responsibilization: On Market-Embedded Morality." *Economy and Society* 37, no. 1 (2008): 1–19.

———. "Capitalism, Governance, and Authority: The Case of Corporate Social Responsibility." *Annual Review of Law and Social Science* 6 (2010): 531–53.

———. "The De-Radicalization of Corporate Social Responsibility." *Critical Sociology* 30, no. 3 (2004): 669–89.

Shellenberger, M., and T. Nordhaus. *The Death of Environmentalism: Global Warming Politics in a Postenvironmental World*. Washington: Environmental Grantmakers Association/The Breakthrough Institute, 2004.

Shrader-Frechette, K.S. *Environmental Justice: Creating Equality, Reclaiming Democracy*. Oxford: Oxford Univeristy Press, 2002.

Siedentrop, L. *Inventing the Individual: The Origins of Western Liberalism*. Cambridge: Belknap, 2014.

Skinner, Q. *Hobbes and Republican Liberty*. Cambridge: Cambridge University Press, 2008.

———. "The Paradoxes of Political Liberty." In *The Tanner Lectures on Human Values*: Harvard University, 1984.

Sklair, L. *The Transnational Capitalist Class*. Oxford: Blackswell, 2001.

Sklar, H. *Reagan, Trilateralism and the Neoliberals*. Boston: South End Press, 1986.

Skocpol, Theda. *Diminished Democracy: From Membership to Management in American Civic Life*. Norman: University of Oklahoma Press, 2003.

———. *States and Social Revolutions: A Comparative Analysis of France, Russia, and China*. Cambridge: Cambridge University Press, 1979.

Skocpol, Theda, and Vanessa Williamson. *The Tea Party and the Remaking of Republican Conservatism*. Oxford: Oxford University Press, 2012.

Smith, Mick. *An Ethics of Place: Radical Ecology, Postmodernity, and Social Theory*. Albany: State University of New York Press, 2001.

Smith, Noah. "The Dark Side of Globalization: Why Seattle's 1999 Protesters Were Right." *Atlantic*, January 6, 2014. http://www.theatlantic.com/business/archive/2014/01/the-dark-side-of-globalization-why-seattles-1999-protesters-were-right/282831/.

Snyder, Gary. *The Practice of the Wild*. San Francisco: Northpoint, 1990.

Steger, Manfred B. *The Rise of the Global Imaginary: Political Ideologies from the French Revolution to the Global War on Terror*. Oxford: Oxford University Press, 2008.

Steger, Manfred B., James Goodman, and Erin K. Wilson. *Justice Globalism: Ideology, Crises, Policy*. Thousand Oaks CA: Sage, 2012.

Stein, Judith. *Pivotal Decade: How the United States Traded Factories for Finance in the Seventies* New Haven, CT: Yale University Press, 2010.

Streeck, Wolfgang. *Buying Time: The Delayed Crisis of Democratic Capitalism*. London: Verso, 2014.

Swyngedouw, Erik, and Nikolas C. Heynen. "Urban Political Ecology, Justice and the Politics of Scale." *Antipode* 35, no. 5 (2003): 898–918.

Tanner, Michael D. *Leviathan on the Right: Leviathan on the Right: How Big Government Conservatism Brought Down the Republican Revolution*. Washington, DC: Cato Institute, 2007.

Taylor, Astra. "Against Activism." *Baffler*, March 2016). http://thebaffler.com/issues/no-30.

Taylor, Blair. "From Alterglobalization to Occupy Wall Street: Neoanarchism and the New Spirit of the Left." *City* 17, no. 6 (2013): 729–47.

———. "Social Ecology in a Neoliberal Age." Unpublished manuscript, 2016.

Taylor, Charles. *Philosophical Arguments*. Cambridge, MA: Harvard University Press, 1995.

———. "Socialism and Weltanschauung." In *The Socialist Idea: A Reappraisal*, edited by Leszek Kolakowski and Stuart Hampshire, 45–58. London: Weidenfeld and Nicolson, 1974.

———. *Sources of the Self: The Making of the Modern Identity*. Cambridge, MA: Harvard University Press, 1989.

Taylor, Dorceta E. "The Rise of the Environmental Justice Paradigm: Injustice Framing and the Social Construction of Environmental Discourses." *American Behavioral Scientist* 43, no. 4 (2000): 508–80.

TEEB. *The Economics of Ecosystems and Biodiversity: Ecological and Economic Foundations*, edited by Pushpam Kumar. London and Washington: Earthscan, 2010.

Tobias, Michael, ed. *Deep Ecology*, 2nd ed. San Diego: Pfeiffer, 1985.

Tokar, Brian. *The Green Alternative: Creating an Ecological Future*, 2nd ed. San Pedro, CA: R. and E. Miles, 1994.

———. "On Bookchin's Social Ecology and Its Contributions to Social Movements." *Capitalism, Nature, Socialism* 19, no. 1 (2008): 51–66.

———. *Toward Climate Justice: Perspectives on the Climate Crisis and Social Change*. Rev. ed. Porsgrunn, Norway: New Compass, 2014.

Torgerson, Douglas. "Democracy through Policy Discourse." In *Deliberative Policy Analysis: Understanding Governance in the Network Society*, edited by Maarten Hajer and Hendrik Wagenaar, 113–38. Cambridge: Cambridge University Press, 2003.

Trachtenberg, Zev. "The Anthropocene, Ethics, and the Nature of Nature." *Telos*, no. 172 (2015): 38–58.

Tully, James. *Democracy and Civic* Freedom, vol. 1 of *Public Philosophy in a New Key*. Cambridge: Cambridge University Press, 2009.

———. *Imperialism and Civic Freedom*, vol. 2 of *Public Philosophy in a New Key*. Cambridge: Cambridge University Press, 2009.

Turner, James Morton. "'The Specter of Environmentalism': Wilderness, Environmental Politics, and the Evolution of the New Right." *Journal of American History* 96, no. 1 (2009): 123–48.

Turner, Bryan S. "Outline of a Theory of Citizenship." In *Citizenship: Critical Concepts*, edited by Bryan S. Turner and Peter Hamilton, 199–226. London: Routledge, [1990] 1994.

Turner, Bryan S. "The Erosion of Citizenship." *British Journal of Sociology* 52, no. 2 (2001): 189–209.

United Nations General Assembly. *The Future We Want*. New York: United Nations General Assembly, 2012, available at https://sustainabledevelopment.un.org/futurewewant.html.

United Nations Global Compact, "Address Environmental Risks and Leverage Opportunities," n.d. https://www.unglobalcompact.org/what-is-gc/our-work/environment

———. "How Business Can Support Development." https://www.unglobalcompact.org/Issues/partnerships/how_business_can_support_development.html.

Vaïsse, Justin. *Neoconservatism: The Biography of a Movement*. Translated by Arthur Goldhammer. Cambridge, MA: Belknap Press of Harvard University Press, 2010.

Van Horn, Rob, and Philip Mirowski. "The Rise of the Chicago School of Economics and the Birth of Neoliberalism." In *The Road from Mont Pelerin: The Making of the Neoliberal Thought Collective*, edited by Philip Mirowski and Dieter Plehwe, 138–78. Cambridge, MA: Harvard University Press, 2009.

Vernon, Raymond, ed. *The Oil Crisis*. New York: Norton, 1976.

Vig, Norman J. "Presidential Powers and Environmental Policy." In *Environmental Policy: New Directions for the Twenty-First Century*, 8th ed., edited by Norman J. Vig and Michael E. Kraft, 84–108. Washington, DC: CQ Press, 2013.

Vig, Norman J., and Michael E. Kraft. *Environmental Policy: New Directions for the Twenty-First Century*, 8th ed. Washington: CQ Press, 2013.

Vogel, David. *Fluctuating Fortunes: The Political Power of Business in America*. New York: Basic Books, 1989.

———. *The Market for Virtue: The Potential and Limits of Corporate Social Responsibility*. Washington, DC: Brookings Institution Press, 2005.

Vogel, Steven. *Against Nature: The Concept of Nature in Critical Theory*. Albany: State University of New York Press, 1996.

Weale, Albert. *The New Politics of Pollution*. Manchester: Manchester University Press, 1992.

Wenz, Peter S. *Environmental Justice*. Albany: State University of New York Press, 1998.

Wheen, Francis. *How Mumbo-Jumbo Conquered the World: A Short History of Modern Delusions*. New York: Public Affairs, 2005.

Wickberg, Daniel. "Waiting for My Man," review of *The Age of the Crisis of Man*, by Mark Greif. *U.S. Intellectual History Blog*, May 13, 2015. https://s-usih.org/2015/05/waiting-for-my-man/.

Williams, Bernard. *Essays and Reviews, 1959–2002*. Princeton, NJ: Princeton University Press, 2014.

———. *Ethics and the Limits of Philosophy*. Cambridge, MA: Harvard University Press, 1985.

———. *In the Beginning Was the Deed: Realism and Moralism in Political Argument*. Edited by Geoffrey Hawthorn. Princeton, NJ: Princeton University Press, 2005.

———. *Truth and Truthfulness: An Essay in Genealogy*. Princeton, NJ: Princeton University Press, 2002.

Wood, Ellen Meiksins. "The Separation of the Economic and the Political in Capitalism." *New Left Review*, no. 127 (1981): 66–95.

Worster, Donald. *Nature's Economy: A History of Ecological Ideas*, 2nd ed. Cambridge: Cambridge University Press, 1994.

Wright, Christopher, and Daniel Nyberg. *Climate Change, Capitalism, and Corporations: Processes of Creative Self-Destruction*. Cambridge: Cambridge University Press, 2015.

Yates, Luke. "Rethinking Prefiguration: Alternatives, Micropolitics and Goals in Social Movements." *Social Movement Studies* 14, no. 1 (2015): 1–21.

Yuhas, Alan. "Protesters Put Bill Clinton on the Defensive at Hilary Rally in Philadelphia." *Guardian*, April 7, 2016. https://www.theguardian.com/us-news/2016/apr/07/bill-clinton-protesters-hillary-rally-philadelphia-incarceration.

Zadek, Simon. *The Civil Corporation: The New Economy of Corporate Citizenship*. London: Earthscan, 2001.

Zaretsky, Eli. *Why America Needs a Left: A Historical Argument*. Cambridge, UK: Polity, 2012.

Zerilli, Linda M. G. *A Democratic Theory of Judgment*. Chicago: Chicago University Press, 2016.

Žižek, Slavoj. *In Defense of Lost Causes*. London: Verso, 2009.

———. "Resistance Is Surrender." *London Review of Books*, November 15, 2007.

Zwick, Steve. "Unpacking Warsaw, Part Two: Recognizing the Landscape Reality." News Articles, Ecosystem Marketplace, December 5, 2013. http://www.ecosystemmarketplace.com/articles/em-unpacking-warsaw-part-two-em-recognizing-the-landscape-reality/.

Index